BIOTECHNOLOGY
AND INDUSTRY

MAINTAINING CULTURES FOR BIOTECHNOLOGY AND INDUSTRY

EDITED BY

Jennie C. Hunter-Cevera

Center for Environmental Biotechnology
Lawrence Berkeley Laboratory
Berkeley, California

Angela Belt

Blue Sky Research Service
Sonora, California

ACADEMIC PRESS

San Diego   New York   Boston
London   Sydney   Tokyo   Toronto

Copyright © 1996 by ACADEMIC PRESS, INC.

Academic Press, Inc.
A Division of Harcourt Brace & Company
525 B Street, Suite 1900, San Diego, California 92101-4495

United Kingdom Edition published by
Academic Press Limited
24-28 Oval Road, London NW1 7DX

Library of Congress Cataloging-in-Publication Data

Maintaining cultures for biotechnology and industry / edited by Jennie
  C. Hunter-Cevera, Angela Belt
          p.   cm.
  Includes bibliographical references and index.
  ISBN 0-12-361946-7 (pbk.
  : alk. paper)
    1. Culture media (Biology) 2. Biotechnology.   I. Hunter-Cevera,
  Jennie C. II. Belt, Angela.
  TP248.25.C44M35   1995
  660'.6--dc20   95-30594
                                                         CIP

PRINTED IN THE UNITED STATES OF AMERICA
96   97   98   99   00   01   BC   9   8   7   6   5   4   3   2   1

Contents

CHAPTER 3
Algae
Robert A. Anderson

CHAPTER 4
Preservation and Maintenance of Eubacteria
L. K. Nakamura

CHAPTER 5

Actinomycetes
Alma Dietz and Sara A. Currie

CHAPTER 6

Fungi
David Smith and Jacqueline Kolkowski

CHAPTER 9
Human and Animal Viruses
Judy A. Beeler

CHAPTER 10
Plant Germplasm
H. R. Owen

List of Contributors

Numbers in parentheses indicate the pages on which the authors' contributions begin.

Robert A. Anderson (29), Bigelow Laboratory for Ocean Sciences, Provasoli-Guillard Center for Culture of Marine Phytoplankton, West Boothbay Harbor, Maine 04575

Judy A. Beeler (179), Division of Viral Products, Federal Drug Administration, Bethesda, Maryland 20852

Angela Belt (251), Blue Sky Research Service, Sonora, California 95370

Sara Currie (85), Roselle, New Jersey 07203

Alma Dietz (85), Microtax, Kalamazoo, Michigan 49006

Albert P. Halluin (1), Pennie & Edmonds, Menlo Park, California 94025

Rose Hammond (229), Molecular Plant Pathology Laboratory, U.S. Department of Agriculture, BARC-West, Beltsville, Maryland 20705

J. Hammond (229), Floral and Nursery Plants Research Unit, U.S. National Arboretum, BARC-West, Beltsville, Maryland 20705

T. Hasegawa (15), Department of Fermentation Technology, Faculty of Engineering, Hiroshima University, Higashi-Hiroshima 239, Japan

Robert J. Hay (161), American Type Culture Collection, Rockville, Maryland 20852

Jacqueline Kolkowski (101), International Mycological Institute, Egham Surrey TW20 9TY, United Kingdom

L. K. Nakamura (65), U.S. Department of Agriculture, Midwest Area National Center, Peoria, Illinois 61604

Thomas A. Nerad (133), American Type Culture Collection, Rockville, Maryland 20852

H. R. Owen (197), Department of Botany, Eastern Illinois University, Charleston, Illinois 61920

Ellen M. Simon (133), Department of Ecology, Ethology, and Evolution, University of Illinois at Urbana-Champaign, Urbana, Illinois 61801

David Smith (101), International Mycological Institute, Egham Surrey TW20 9TY, United Kingdom

Preface

One of the most valuable assets in any biological research organization, be it industry, academia, or government, is the culture repository. Whether one works independently in the laboratory, maintaining cultures for limited use, or as part of a larger, company-wide effort, the culture collection represents the heart of research efforts from basic and applied to development and manufacturing. Many companies have one or more well-staffed and equipped collections that serve different functions such as providing reference strains, assay strains, strains for screening purposes, and production strains. On the other hand, some small biotechnology companies or departments within universities may not be so well funded and supplied with appropriate equipment for maintaining these resources. In biotechnology companies and institutes, such as pharmaceutical, agrobiological, and biocatalysis, the research can often be quite focused or diverse in terms of "biodiversity" and include wild-type parent or reference strains and/or genetically engineered strains.

Maintaining culture traits, viable titers, and plasmids, in the case of recombinant strains, is important for stability of production strains, purity, and reproducibility of research. Most of the staff usually associated with culture collections are highly trained individuals who have the experience and expertise to identify and properly store or maintain cultures over long periods of time. Sometimes support is limited for culture collections in terms of lab space, staff, and monetary assets. This, in turn, may limit the method used to characterize, maintain, and preserve cultures. However, when a culture problem arises that may cost the company financial loss, major attention is usually given to the curator and his or her staff. The value that a curator and collection provides to research efforts is immeasurable and seldom fully appreciated.

With this in mind, we felt that a book addressing the various needs of maintaining and preserving the biodiversity within large and small biotechnological companies and institutions was necessary. In addition to the literature on maintaining cultures, techniques that have been passed on to graduate students or research associates by word of mouth and never published are included. The methods described are the product of the hands-on experience of our authors who have all worked with cultures in laboratories for many years. The readers will find it interesting to learn that there is more than one way to maintain cultures and that choices can be made based on equipment and budget considerations. It was also our intent that both students and experts in the field of maintaining cultures will find valuable use in the details given for simple preservation methods and in the references included, which provide additional information on maintaining each group of organisms addressed.

The authors have prepared a general introduction including discussions of biodiversity from a taxonomic or phylogenic point of view, an overview of classification, examples of economic or industrial importance, and information on characterizing cultures. Lists of culture repositories where strains can be acquired or deposited and references for more additional reading follow most chapters.

The book is organized into twelve chapters that cover the maintenance and preservation of algae, eubacteria, actinomycetes, fungi, protozoa, animal cells in culture, human and animal viruses, plant germplasm, plant viruses, and viroids. We begin with an introductory chapter on the importance of culture collections from a patent attorney's point of view. Also included in this chapter is a list of recognized culture collections for depositing patent strains under the Budapest Treaty. Subsequent chapters summarize the history and evolution of culture maintenance, and preservation techniques are described in detail. The final chapter suggests ways to evaluate and characterize cultures used for biotechnology and related industries.

Overall the book takes the mystery and fear out of maintaining and preserving cultures for biotechnology and encourages those both young and old to take pride in the job of being the "keeper" of the cultures. It can also be a valuable educational tool for managers to better understand the real value of culture collections and what is needed by curators to best maintain stability of cultures and reproducibility of desired traits for commercial profit. We believe this book will be a fine addition to the reference library of anyone interested in maintaining cultures for research.

Jennie C. Hunter-Cevera
Angela Belt

CHAPTER 1

The Biological Deposition Requirement

Albert P. Halluin*

Introduction

As biotechnology ascends in significance in the overall scheme of modern innovation, mechanisms governing society's interaction with innovation have been forced to adapt to its particular characteristics. Given the enormous potential of biotechnology to benefit society, it is incumbent that industry, government, and academia continue to accommodate the special needs of this emerging field of scientific endeavor.

The establishment of depositories for biological materials and the development of specific sample preservation techniques have evolved in response to heightened interest in biological investigation. And as society seeks to ensure unfettered access to the perishable fruits of this research, the deposition and maintenance of organisms assume an even greater importance. By making available samples of established origin and quality, society may dramatically increase the efficiency of investigation. Existing cell lines and genetic materials may be made uniformly available to all researchers, providing a common starting point as well as a basis for comparison of the work of different groups.

*Partner, Pennie & Edmonds, Menlo Park, California. The author expresses his gratitude for the assistance of Kent Tobin in the preparation of this chapter. Mr. Tobin holds a Bachelor of Science Degree in chemistry from the University of California, San Diego, and is a graduate of the University of California, Hastings College of Law in San Francisco.

At present there exist international depositories for biological entities, such as the American Type Culture Collection in Rockville, Maryland, the Fermentation Research Institute in Japan, the Centraal Bureau voor Schimmelcultures in The Netherlands, and the German Institute for Micro-organisms (Halluin, 1982). Given the enormous variety of materials implicated in future biotechnological research, from unincorporated genetic material to antibodies to multicellular organisms, there is great need for uniform, detailed procedures regarding the care of such samples.

The purpose of this volume is to provide such guidance. By way of introduction, however, it is perhaps first useful to examine the motivation for deposition. In particular, the focus will be on the role played by the law in mandating public availability of biological entities that are the subject matter of a patent.

As of the writing of this chapter, the status of the requirement for making biological patent depositions is uncertain. Recent decisions handed down by the Court of Appeals for the Federal Circuit have significantly altered the long-standing policy favoring deposition (Halluin and Wegner, 1991; Halluin, 1992). The impact of this shift in philosophy has yet to be recognized fully, and cannot be underestimated given the importance of pharmaceutical research to modern society.

Concerns Other Than Patentability

Aside from the patentability concerns discussed below, there are many excellent reasons for maintaining depositories of available biological material. These include (1) the logistical convenience of a centralized storage location, (2) the security of maintaining backup samples should the working samples perish by mistake, (3) the advantage in safety conferred by ensuring that all who seek to obtain a particular material are adequately informed of its potential environmental and health risks, (4) the assured consistency in handling and preservation of samples within a particular depository, and (5) an enhanced ability to monitor the possessor and location of a particular sample at any given time (Hunter *et al.*, 1986).

This last rationale is especially important to the accurate assessment of issues such as licensing, trade secrecy, and patent infringement of a given composition, especially those provided by outside sources. The weight of these concerns will vary according to the purpose of the depository, but all can be envisioned as significant at some point in the life span of an average depository.

An Introduction to the Patent System

The patent system has its origins in the United States Constitution of 1789. Article I, Section 8, Clause 8 provides Congress with the power "[t]o promote the progress of science and the useful arts, by securing for limited times to authors and inventors the exclusive right their respective writings and discoveries." This intent has been explicitly elaborated and codified by United States Code, Title 35, whose provisions establish the Patent and Trademark Office as a division of the Department of Commerce.[1]

The effect of granting a patent is to assure the inventor a 17-year exclusive right to make, use, or sell an invention. Issuance of a United States patent is not a guarantee of such a monopoly; it merely creates a presumption of validity that may be challenged in court by a competitor.[2]

As with all legal doctrines, patent law is shaped and governed by underlying social policies, whose recognition is indispensable to understanding the operation of the law. There is commonly said to be one major policy disfavoring the granting of patents. This policy is the deeply rooted American antipathy to governmentally created and enforced monopolies. Thomas Jefferson, a noted inventor in his own right, strongly advocated the limiting of monopolies wherever possible, going even so far as to propose a provision in the Bill of Rights limiting the duration of any monopoly.[3] Although this provision was never formally adopted, the traditional American reluctance to interfere with the operation of a free market economy endures to the present day. This aversion is evidenced by the strictly limited duration of the patent term to 17 years or 20 years from filing.

Opposing the action of the antipathy to monopolies are two main policies. First is the incentive/reward rationale. This postulates that the granting of a period of exclusive sale is intrinsically beneficial to human creativity, because it provides an economic incentive to the inventor in the form of a monopoly over the sale of the product. This policy operates not only to reward the inventor, but also to prevent competitors from

---

[1]35 USCA §1. Establishment, reads in pertinent part:

The Patent and Trademark Office shall continue as an office in the department of Commerce, where records, books, drawings, specifications, and other papers and things pertaining to patents and trademark registrations shall be kept and preserved, except as otherwise provided by law.

[2]The legal effect of a presumption is to place the burden of persuasion on the party challenging validity. The judge or jury will commence their analysis from the proposition that the patent is in fact valid. The party opposing the patent must then provide evidence stating otherwise (Harmon, 1988).

[3]See *Graham v. John Deere Co.*, 383 U.S. 1, 8–9 (1966).

reproducing the invention and thereby profiting without having invested in the development of the invention (Menges and Nelson, 1990). Some commentators have questioned this policy, claiming that trade secret protection coupled with the intrinsic advantage conferred by a head start is sufficient to ensure adequate compensation for innovation (Kitch, 1977).

A second policy favoring the granting of patents is that of disclosure. Title 35 of the United States Code requires that any invention awarded a patent will be publicly disclosed in sufficient detail so as to enable a practitioner of average skill in the art to reproduce the invention without undue experimentation. Disclosure is viewed as favorable to creativity by allowing the general community to have access to the invention and thereby facilitating subsequent improvement (Menges and Nelson, 1990).

The formal manner by which the patent system seeks to ensure the full disclosure of inventions is spelled out in the first paragraph of Section 112 of Title 35 of the United States Code.

> The specification[4] shall contain a written description of the invention, and the manner and process of making and using it, in such full, clear, concise and exact terms so as to enable any person skilled in the art to which it pertains, or with which it is most nearly connected, to make and use the same, and shall set forth the best mode contemplated by the inventor of carrying out his invention.

Three distinct requirements for patentability are stated in this paragraph. First is the written description requirement, whose purpose is to ensure that the inventor's claims are adequately documented and supported by the specification. Second is the enablement requirement, whose purpose is to mandate that the disclosure empowers the ordinary artisan to make and use the invention, given only the written description. Third, the "best mode" requirement exists to ensure that the public benefits fully from the patentee's knowledge of his best work, which cannot be withheld as a trade secret (Chisum, 1992).

The Deposition Requirement

The patent system has a long history of requiring the submission of materials in addition to a written description. The Patent Act of 1793 required specimens of ingredients of composition-of-matter claims to be submitted to the Patent Office. Section 6 of the Patent Act of 1836 required

---

[4]The two most important components of any patent application are the claims and the specification. The former formally states the limits and content of those features to which legal title will be asserted. The latter refers to the comprehensive overall description of the invention and is analogous in form and content to a scientific article (Goldstein, 1990).

the inventor to furnish a model of the invention in order to exhibit "advantageously" its features. By 1880, however, difficulties with the storage and display of models led to the waiver of these requirements. To this day, the Patent and Trademark Office retains the right to require the submission of models or specimens along with the written specification and claims.[5]

The current debate surrounding the deposition of biological materials is of a somewhat different origin, however. Rather than requiring deposition to allow the examiners to reach a more fully informed determination of patentability, deposition is regarded as a prerequisite for full public disclosure of the invention, as codified by the enablement and best mode doctrines discussed above.

Proponents of deposition argue that, without public availability of the physical item itself, written disclosure alone is insufficient to enable an artisan of ordinary skill to practice the invention. Proponents of deposition also assert that a mere writing does not constitute the inventor's most efficient and valuable means of practicing the invention. Rather, the organism or genetic material must be made available to the public, thereby conveying physical possession of materials unable to be duplicated chemically or otherwise.

The above arguments reflect an underlying fear that inventors will consciously forego deposition of cell lines so as to obtain trade secrecy protection in conjunction with their patent protection. This behavior would pose a grave threat to the continuing free flow of information within the scientific community, and is perceived as an intolerable result by many commentators (Halluin, 1992).

Until fairly recently patent law set a relatively high threshold for what materials were sufficiently accessible by the public to render unnecessary their forced availability. In 1975, the standard embraced by the highest patent court (the United States Court of Customs and Patent Appeals, now replaced by the Court of Appeals for the Federal Circuit, hereafter "Federal Circuit") stated that for purposes of enablement, a biological material need not be deposited if it was known and readily available to the public.[6] Any unknown or new compound or organism would thus be required to be deposited. In 1991, however, the standard was seemingly relaxed for certain types of biotech-type inventions whereby the Federal

[5] 35 USCA §114. Models, specimens

The Commissioner may require the applicant to furnish a model of convenient size to exhibit advantageously the several parts of his invention.

When the invention relates to a composition of matter, the Commissioner may require the applicant to furnish specimens or ingredients for the purpose of inspection or experiment.

[6] *Feldman v. Aunstrup,* 517 F. 2d 1351, 1354 (CCPA 1975).

Circuit held that no deposition was mandated to satisfy the best mode requirement, so long as "the cells can be prepared without undue experimentation from known materials, based on the description in the patent specification.[7]

The "undue experimentation" language of the latter decision is significant in that it opens the door for the court to interpret subjectively how much experimentation is "undue." This approach must be contrasted with the earlier decision mandating deposition if the organism were "new," a threshold issue that could often be objectively determined given the facts of the case and the state of the prior art.

The Current Debate

The cases discussed above indicate that the Federal Circuit will attempt to relax the deposition requirement under some circumstances. To understand why this may be happening, it is important to first recognize certain jurisprudential principles that interact to shape the law.

One extremely important motivating force is a judicial desire that the law apply uniformly and consistently to the innumerable potential fact situations it will be called on to resolve. The law must remain flexible enough to address problems that arise within its jurisdiction without becoming enmeshed in artificial distinctions and classifications.

It is the expressed intent of the Federal Circuit that "[t]he law must be the same for all patents and types of inventions. A level playing ground for the marketplace of ideas is as necessary for technological innovation as it is for politics and social policy."[8] In the realm of patent law, no discipline other than biology requires that a physical embodiment of the invention be publicly available, regardless of the complexity of the subject matter at issue. To carve out a particular exception for one science seemingly violates the uniformity principle mandated by the Federal Circuit.

It is important to emphasize that the concern for simplicity is not merely the product of aesthetic intellectual concerns. Rather, the presence of extra-

---

[7]*Amgen v. Chugai Pharmaceutical Co., Ltd.*, 927 F. 2d 1200, 1211 (Fed. Cir. 1991); see also *Scripps Clinic and Research Foundation v. Genentech, Inc.*, 927 F. 2d 1565 (Fed. Cir. 1991). The Scripps patent addressed the isolation of a monoclonal antibody. The Federal Circuit determined that "Genentech's argument is primarily that because of the laborious nature of the process of screening monoclonal antibodies, the inventors should have voluntarily placed in depository and made available to the public the antibody . . . [a]lthough Genentech suggests that Scripps should have made a deposit voluntarily, failure to do so can not constitute legal or factual basic for patent invalidity." *Id.* at 1579. The Scripps decision is thus evidence of the Federal Circuit's reluctance to require a deposit on the basis of "undue experimentation".

[8]*Panduit Corp. v. Dennison Mfg. Co.*, 1 USPQ2d 1593, 1602 (Fed Cir. 1987).

neous categories complicates the legal analysis. Unnecessary distinctions carry with them the pragmatic danger of judicial inefficiency. The court may allocate its time and energy into classifying the invention according to some artificial construct, rather than addressing the fundamental issues posed by the facts of the case. Such an analysis may lead to grossly inequitable results and is to be avoided.

However, research in the biological sciences may carry with it such unique and significant characteristics that the accommodating exception is justified.

> As the science of biotechnology matures the need for special accommodation, such as the deposit of cell lines or microorganisms, may diminish; but there remains the body of law and practice on the need for sufficient disclosure, including experimental data when appropriate, that reasonably support the scope of the requested claims. That law relates to the sufficiency of the description of the claimed invention, and if not satisfied by the deposit, must independently meet the requirements of Section 112.[9]

One such difference stems from the very nature of the entities studied by biology. In no other field of research can the products of investigation reproduce themselves without assistance of having the actual specimen in hand. In fact, the rules promulgated by the United States Patent and Trademark Office essentially define biological materials using this property.[10]

Also of great consequence is the enormous complexity of living systems, and the limited ability of modern science to deduce the structures and mechanisms governing their operation. These inescapable limitations lead to an amount of uncertainty not present in other areas. Manifestations of this characteristic include (1) the impossibility of quantifying minute errors in replication of genetic material millions of base pairs long, (2) the impossibility of reproducing properties of particular antibodies, even if the original procedure is followed precisely, and (3) the difficulty of isolating certain organisms and natural products from their surrounding environment (Hampar, 1985). The existence of this underlying uncertainty is so pronounced that it has been explicitly recognized by the courts:

[9]*In re Wands,* 8 USPQ2d 1400, 1408 (Fed. Cir. 1988) concurring opinion of J. Newman.
[10]Title 37 Federal Code of Regulations Section 1.801. Biological material.
For the purposes of these regulations pertaining to the deposit of biological material for purposes of patents for inventions under 35 USC section 101, the term biological material shall include material that is capable of self-replication either directly or indirectly . . . [v]iruses, vectors, cell organelles and other non-living material existing in and reproducible from a living cell may be deposited by deposit of the host cell capable of reproducing the non-living material.

> In cases involving predictable factors, such as mechanical or electrical elements
> . . . once imagined, other embodiments can be made without difficulty and their
> performance characteristics predicted by resort to known scientific laws. In cases
> involving unpredictable factors, such as most chemical reactions and physiological
> activity, the scope of enablement obviously varies inversely with the degree of
> unpredictability of the factors involved.[11]

This uncertainty brings into question the ability of a mere written disclosure to satisfy either the best mode or enablement requirements. Recent cases have held exact duplication of results not to be necessary to satisfy these doctrines.[12] Rather, the disclosure must merely be "adequate" to reveal the invention and the best mode of practicing it.[13] The subjectivity of such a standard is not conducive to predictability of outcomes. Thus, the inquiry remains whether the properties and structure of a material are able to be reproduced by an artisan of ordinary skill in the art, relying solely on a written description. If satisfied, the written disclosure has met the underlying public interest by providing unrestricted accessibility to patented inventions.

Other concerns operate to favor the availability of samples. There is substantial convenience to all researchers in having the biological specimen immediately available.

> The value of a bioculture deposit lies in the fact that scientists know precisely what
> they are working with. Their experiments and results are therefore comparable
> with those of other scientists who have worked on the same subject. The pioneering
> work of others can be verified, replicated, and built upon . . . [d]eposits thus save
> taxpayers, who fund millions of dollars worth of bioresearch every year, countless
> dollars, because universities and foundations who receive federal funds do not have
> to spend months or years trying to find the starting point. Without deposits, much
> research would become prohibitively expensive and never be undertaken.[14]

Although not traceable to the best mode or enablement requirements, such an increase in the efficiency of research is extremely desirable, and should

---

[11]*In re Fisher,* 166 USPQ 18, 24 (CCPA 1970).

[12]See *Christison v. Colt Industries Operating Corp.,* 870 F. 2d 1292 (7th Cir. 1989), where the 7th Circuit held that a gun manufacturer need not disclose the blueprints of the invention in order to satisfy the best mode requirement. *Id.* at 1302. It is interesting to note the distinction articulated in Fisher: here, the exact duplication of mechanical and electrical inventions is not necessarily the same as that of biological entities. Differences resulting from the uncertainty inherent in biological systems are not merely the product of variances in tolerances spelled out in a blueprint. Rather, the intrinsic variability of biological activity is so profound as to affect the underlying structure of the invention itself. The description of the invention is not merely more "fuzzy"; it may disclose a different invention altogether. In this context, the distinction is a crucial one.

[13]*Amgen v. Chugai Pharmaceutical Co., Ltd.,* 927 F. 2d 1200, 1212 (Fed. Cir. 1991).

[14]Brief Amicus Curiae of the American Type Tissue Culture Collection at 7, *Genetics Institute Inc., and Chugai, Pharmaceutical Co., Inc., v. Amgen, Inc.,* 927 F.2d 1200 (Fed. Cir. 1991).

be taken into account when deciding on the overall merits of mandatory deposition.

How the courts will ultimately resolve the tension between the desire for uniformity and the special needs of biotechnology is unknown. Given the immense potential societal value of biopharmaceutical research, great efforts will be devoted to deciding this issue so as to provide a maximum impetus to further research.

In discussing the patentability of inventions, however, it is important to remember that the United States is only one component in the global economy. The patent laws of other countries often differ substantially and must be considered in any discussion of the legal aspects of biological deposition.

Global Deposition Requirements

In recent years the practice of depositing living materials in conjunction with the filing of a patent application has become relatively common throughout the world. In 1977, the Budapest Treaty on the International Recognition of the Deposit of Micro-Organisms for the Purposes of Patent Procedure was adopted to facilitate this practice on an international scale. The Treaty mandates member nations requiring or recommending deposition to recognize deposits made in any of a specific goal network of depositories. These are listed at the end of this chapter. The provisions of the Treaty serve to simplify the filing of foreign applications, and also to set forth uniform procedures for making a deposit. As of 1989, the United States, Japan, South Korea, France, Germany, the United Kingdom, and most other industrialized nations were signatories.

Presently the patent systems of the United States, Germany, Japan, the United Kingdom, South Korea, the Peoples Republic of China, and the European Patent Convention have some provision calling for the deposition of biological materials not otherwise accessible to the public. The particular applicable standard will of course vary, but the conclusion is clear: any inventor seeking worldwide patent protection for his or her biological invention is advised to deposit in at least one of the depositories recognized by the Budapest Treaty. Failure to do so runs the risk of rejection on grounds of inadequate disclosure.

Although under attack in some circles, the deposition requirement continues to play a key role in the issuance of valid patents both domestically and throughout the world. The ubiquity of this requirement ensures that regardless of the eventual outcome of the present domestic debate in the United States, mandatory deposition will continue to be a consideration in patent practice for years to come, making this volume an invaluable resource not only for scientists, but for patent agents and attorneys as well.

International Depository Authorities under the Budapest Treaty as of August, 1991[15]

Agricultural Research Service Culture Collection (NRRL)
1815 North University St.
Peoria, IL 61604

American Type Culture Collection (ATCC)
12301 Parklawn Dr.
Rockville, MD 20852

Australian Government Analytical Laboratories (AGAL)
New South Wales Regional Laboratory
1 Suakin St., Pymble
N.S.W. 2073
Australia

CAB International Mycological Institute (CAB IMI) Ferry Lane
Kew, Surrey TW9 3AF
United Kingdom

Centraal Bureau voor Schimmelcultures (CBS) Oosterstraat 1
P.O. Box 273
3740 AG Baarn
The Netherlands

Collection Nationale de Cultures de Micro-organismes (CNCM)
Institute Pasteur
28, Rue du Docteur Roux
75724 Paris Cedex 15
France

Culture Collection of Algae and Protozoa (CCAP)
Institute of Freshwater Ecology
The Windmere Laboratory
Far Sawrey, Ambleside, Cumbria LA22 OLP
United Kingdom

Culture Collection of Algae and Protozoa (CCAP)
Dunstaffnage Marine Laboratory
P.O. Box 3
Oban, Argyll PA34 4AD, Scotland
United Kingdom

[15]Source: Manual for the Handling of Applications for Patents, Designs and Trademarks throughout the World, Manual Industrial Property B.V., Amsterdam.

Deutsche Sammlung von Mikrooganismen und Zellkulturen GmbH (DSM)
Mascheroder Weg 1b
W-3300 Braunschweig
Germany

European Collection for Animal Cell Cultures (ECACC)
Porton Down
Salisbury, Wiltshire SP4 OJG
United Kingdom

Fermentation Research Institute (FRI)
I-3, Higashi 1-chome
Tsukuba-shi, Ibarkai-ken, 305
Japan

IMET—Nationale Sammlung von Midroogransismen
IMET—Hinterlegungsstelle
Beutenbergstrasse 11
O-6900 Jena
Germany

Institute of Microorganism Biochemistry and Physiology of the USSR Academy of
Science (IBFM)
Collection of Microorganisms
Pushchino-na-Oke
USSR-142292, Moscow Region
Soviet Union

Korean Collection for Type Cultures (KCTC)
Genetic Engineering Research Institute
Korea Institute of Science and Technology
05-333, 1 Oun-Dong
Yusong-Gu, Taejon
Republic of Korea

Korean Culture Center of Microorganisms (KCCM) College of Engineering
Yonsei University
Sodaemun-Gu, Seoul 120-749
Republic of Korea

National Bank for Industrial Microorganisms and Cell Cultures (NBIMCC)
125 Lenin Blvd.
Block 2, Sofia
Bulgaria

National Collection of Agricultural and Industrial Microorganism (NCAIM)
Department of Microbiology
University of Horticulture Somloi ut 14-16
H-1118 Budapest
Hungary

National Collection of Food Bacteria (NCFB)
AFRC Institute of Food Research
Reading Laboratory
Shinfield, Reading RG2 9AT
United Kingdom

National Collection of Type Cultures (NCTC) Central Public Health Laboratory
175 Colindale Avenue
London NW9 5HT
United Kingdom

National Collection of Yeast Cultures (NCYC)
AFRC Institute of Food Research
Norwich Laboratory
Colney Lane
Norwich NR4 7UA
United Kingdom

National Collections of Industrial and Marine Bacteria Ltd. (NCIMB)
23 St. Machar Dr.
Aberdeen AB2 1RY, Scotland
United Kingdom

USSR Research Institute for Antibiotics of the USSR Ministry of the Medical and
Microbiological Industry (VNIIA)
Collection of Microorganisms
Nagatinskaya Street 3-a
USSR-113105, Moscow
Soviet Union

USSR Research Institute for Genetics and Industrial Microorganism Breeding of the
USSR Ministry of the Medical and Microbiological Industry (VNIIGIMI)
Collection of Microorganisms
Dorozhnaya St. 8
USSR-113545, Moscow
Soviet Union

References

Chisum, D. (1992). "Patents: A Treatise on the Law of Patentability, Validity and Infringe-
    ment," p. 7–7. Matthew Bender and Co., New York.
Goldstein, P. (1990). "Copyright, Patent, Trademark, and Related State Doctrines," 3rd ed.,
    pp. 451–453. Foundation Press, Westbury, NY.
Halluin, A. P. (1982). Patenting the results of genetic engineering research: An overview.
    Banbury Rep. **10,** 87.
Halluin, A. P. (1992). Withholding patented biocultures from public depositories: Will the
    trend harm medical progress? *HealthSpan* **9** (8), 5–6.
Halluin, A. P., and Wegner, H. C. (1991). *Amgen* and *Scrips:* The new biotechnology practice.
    Biotechnol. Law Rep. **10**(3), 206.

Hampar, B. (1985). Patenting of recombinant DNA technology: The deposit requirement. *J. Patent Trademark Off. Soc.* **67,** 585.

Harmon, R. (1988). "Patents and the Federal Circuit," p. 12. Bureau of National Affairs, Washington, DC.

Hunter, J. C., Belt, A., and Halluin, A. P. (1986). Guidelines for establishing a culture collection within a biotechnology company. *Trends Biotechnol.* **4,** 5–6.

Kitch, E. (1977). The nature and function of the patent system. *J. Law Econ.* **20,** 276–278.

Menges, R., and Nelson, R. (1990). On the complex economics of patent scope. *Columbia Law Rev.* **90,** 880–891.

CHAPTER 2

History and Evolution
of Culture Maintenance
and Preservation Techniques

T. Hasegawa

Historical Review of Maintenance
and Preservation Techniques
for Microorganism Strains

Techniques to maintain and preserve microorganisms have become increasingly important in recent years. Development of such techniques is necessary to ensure the availability of microorganisms for practical use in research and for taxonomic studies. For example, in 1978, a revisionist taxonomic system that placed bacteria and cyanophyta in the Kingdom Prokaryotae was adopted by the International Committee on Systematic Bacteriology; cultured cells, maintained and preserved using various protocols, were the basis for inclusion of named species in this new biological system.

Culture maintenance and preservation techniques have become the center of broad scientific interest since the 1930s, but they originated hundreds of years ago. Long before human beings knew yeasts existed, liquors peculiar and indigenous to each nation or region were being brewed. It is very probable that, based on repeated experiences, ancient brewers developed a technique to use a portion of the brewage as "common seed" for the next brewing. Examples of this are still found in South East Asian countries. In Japan, for over 300 years, there has existed a remarkable profession call "Koji-ya," in which spores of *Aspergillus* are cultured and sold as brewing materials. This is but one of the modes long used for maintenance of microorganisms. It is only more recently that the modern concept of strains and their maintenance has been established on a large scale for applications in industry and research.

Toward the end of the seventeenth century, in the 1670s, van Leewen-hoek (1677), a Hollander, made observations of natural water (rainwater and seawater), as well as fermented beer. By using a coarse, self-made microscope, he found countless small "animalcules" squirming in these liquids and therefore had the honor of being the first to discover the world of microorganisms. There were no further developments in the field of microbiology in the eighteenth century. At the beginning of the nineteenth century, the opinion was still dominant among chemists that the alcoholic accumulation during the brewing process of beer and wine was due to chemical reactions rather than to living organisms. In the 1830s, Cagniard-Latour (1838), of France, discovered that the production of alcohol was due to cellular metabolism of "bacteria," which were later named "sugar bacteria" by Schwann (1837). The following year, Meyen (1838) proposed *Saccharomyces cerevisiae* as the scientific name of these organisms by directly translating their original name in Latin. Twenty years later, there was another milestone in microbiology: the arguments from chemists, against the theory of biological fermentation, were completely refuted by Pasteur (1860). Pasteur also tried various heating methods to exclude influences from airborne microorganisms. His ideas and methods were adopted not only in agriculture but also in medicine, resulting in major advances in surgical techniques and pathogenic microbiology. Pasteur's greatest achievement was to demonstrate that "every type of fermentation was mediated by specific microorganisms." However, solid culture media were not available in those days, and Pasteur failed to isolate pure cultures.

The original solid culture medium was sections of potato or melon. By the time Pasteur demonstrated the presence of microorganisms in the air, bacteriologists had managed to obtain colonial growth from cultured broth inoculum on solidified culture medium, but it was not until the 1870s that colonies were isolated and observed. A well-known microbiologist and plant pathologist, O. Brefeld, first used gelatin to solidify culture medium. He depicted the growth process, from monospore, to hyphal growth, to spore formation, for *Penicillium glaucum*. Robert Koch, the father of medical bacteriolgy, also used sections of potato at first, then when Loeffler (1881) published a nutrient broth, Koch solidified it with gelatin. Perhaps he overlooked the work of Brefeld. The use of agar by Koch brought a considerable improvement to the techniques for isolating bacteria as pure cultures. However, it is reported that Mrs. Hesse, the wife of one of Koch's students, played an important role behind the scenes, by suggesting the use of gelatin as solidifying agent.

A Danish botanist, E. C. Hansen, founded the basis of mycology in the field of brewage, using mainly yeasts. He modified established isolation techniques to improve pure culture methods. In addition, he designed a small chamber, glassed on four sides and on top, to facilitate aseptic procedures. Hansen's sterile box is widely used even today. The sterile box was disinfected with 0.05% solution of corrosive sublimate or 50% alcohol. All other instruments were sterilized by flame treatment or using antiseptic solutions. The pure culture method invented and refined by Hansen for yeast has become the basis of aseptic microbiological techniques.

There are significant differences between microorganisms occurring in nature and those artificially cultured, but our knowledge of these differences is insufficient at present. Moreover, although the strains of microorganisms to be stored for long periods of time, and to be used repeatedly for research and/or production, are pure cultures, they are prone to spontaneous mutations during storage. The necessity to address such problems had already been realized when techniques for culture preservation and maintenance were first developed. In the early days, nutrient broth and malt extract were used for culturing bacteria and fungi, respectively. These cultures were maintained on solid media made with agar or gelatin, incubated at room temperature in the laboratory, and required successive transfers or subcultivations at appropriate intervals. It is possible that the physiological activities recorded initially changed in response to the frequency of the subcultivation and with increasing cellular growth. It was discovered that the solution to this potential problem was to minimize microbial growth and suppress metabolic activities. The first data from long-term storage were reported by Will (1907); various yeast species, freshly cultured and suspended in a sterile 10% sucrose solution, were found to survive in storage for as long as 8 years. However, their physiological activities were significantly reduced and did not recover even after repeated inoculation on fresh medium. Hansen and Lund (1939) recommended suspension of yeasts in a 10% sucrose solution and storage in a cool, dark place, and reported that this method was also applicable to fungi, when grown on nutrient agar slants. Although techniques in the 1930s were no better than that described above, some investigators attempted newer methods. For example, Shackell (1909) studied methods to vacuum-dry live materials, including rabies virus, particularly in the frozen state. Hammer (1911) introduced the first freeze-drying methods for storage of *Escherichia coli, Staphylococcus aureus,* and *Pseudomonas aeruginosa.* Lumiere and Cherrotier (1914) developed a method for storing *Gonococcus* in which the culture medium containing serum was superposed with liquid paraffin. This report did not attract any attention from microbiologists, but as soon as Morton and Pulski (1938) recommended the method, it began to be used for various microorganisms because of its simplicity, without need for special instruments, and its applicability to both culture and effectiveness for long-term storage. As described previously, at the end of the nineteenth century, when the pure culture method was established and research was dependent on using purely isolated strains, the need to maintain microorganisms became essential.

In 1890, Frantisek Král established the world's first collection of bacteria and fungi in Prague (Kocur, 1990). It was a typical example of a private collection in the late nineteenth century. It existed only until 1911. The collection included strains of *Mycobacterium tuberculosis* and *Agrobacterium tumefaciens* (*Bacterium tumefaciens*) offered in lot from R. Koch and E. Smith, respectively. Unfortunately, almost all of the strains were lost. Thereafter, numerous private collections were established in England, Holland, Germany, Japan, and the United States. These collections have been gradually replaced by public culture collections. In Holland, for example, the International Botany Association founded an institute that collected and stored new species of fungi for research investigators. A catalog of collected strains was published in 1906. It was the first step toward the establishment of the Centraal Bureau voor Schimmelcultures (CBS), one of the largest fungal culture collections in the world. In 1899, when the American Bacteriological Society was organized, a bacterial collection for public use was proposed. It was finally established at the National History Museum of New York City in 1911. This was the predecessor of the American Type Culture Collection (ATCC) founded in 1925, which today is one of the largest collections in the world. In Japan, characterized microorganisms were originally maintained in universities, investigational institutes, and private companies. In 1944, at the request of the Cabinet, Takeda Chemical Industries, Ltd. founded the Institute for Aerial Fermentation to collect, store, and distribute useful microorganisms for research on and production of aviation fuel, drugs, and foods. The institute was named the Institute for Fermentation, Osaka (IFO) after World War II and is now the largest culture collection in Japan.

One of the troublesome issues still inherent to maintenance and preservation of microorganisms is the phenomenon of spontaneous mutation occurring among cultured cells. With repeated subcultivations, their viabilities decrease gradually when compared with the parent cultures, finally declining into a continuous and irreversible degeneration. In the case of fungi, for example, differentiation potentials in morphogenesis will be reduced and the physiological activities affected. Hansen and Smith (1932) first analyzed such phenomena and demonstrated that it was due to heterokaryosis in fungi.

In addition to research aimed at understanding mutagenesis during storage, research on the production of antibiotics was initiated in the late 1940s, resulting in gradual changes in preservation techniques. In the interest of "product stability," a number of investigations on storage technique were carried out and the results were published. For example, procedures employing lyophilization and liquid paraffin were effective when used in combination with storage at low temperatures. By using these methods and others, such as carrier-mediated storage, in which microorganisms are adsorbed on soil or sand, as well as freeze storage in deep freezers or liquid nitrogen, storage for several to 10 or 20 years was achieved (Fennel *et al.*, 1950; Hartsell, 1956; Hunt *et al.*, 1958; Hwang, 1960; Jensen, 1961; Smith, 1983).

Use of Microorganisms in Industry

Today, applied microbiology is a science that involves the selective production of various genes. In other words, it purposes are to modify microorganisms by manipulating or inserting genes so that they either (1) produce mutants that do not exhibit normal metabolic pathways or (2) produce substances otherwise never produced by microorganisms. It has been only for about 100 years, however, that we have been able to isolate, in pure culture, the bacteria and molds that produce useful substances for industries and to improve their production through mutagenesis. In addition, when microbial genetics first became available, industrially useful strains were "made over," by cloning, to meet particular requirements. During and after 1945, the discovery of many chemical and radioactive mutagens provided microbiologists with useful means to control or alter genes. Moreover, in the 1940s, reproductive phenomena, such as DNA exchange or conjugation, were found to occur in bacteria, and new genetic phenomena were discovered in fungi (Hansen, 1942). Remarkable progress in microbial genetics and molecular biology has resulted from continuous efforts to better understand these phenomena.

During several years after World War II, the industrialization of antibiotic production changed the industry significantly with respect to productivity and volume of products. For example, in addition to penicillin, which had already been produced during the War, many bioactive substances were discovered to be effective against diseases caused by bacteria and fungi as well as against various malignant tumors. Development of new fermentation methods enabled commercial microbial production of chemicals, including amino acids and nucleotides. The application of gene manipulation techniques increased the production of these industrial products at a higher efficiency. Since 1973, the development of new technologies, including recombinant DNA and cloning, has enabled the introduction of genes of, theoretically, all living things into microorganisms. The possibilities of gene engineering are infinite. Microorganisms now produce fermentation products such as human insulin and growth hormones, as well as high yields of any given molecule of interest.

The dramatic innovations and progress of biotechnology in recent years have changed our view of life. Today, tissue culture methods, one of the fundamental techniques of biotechnology, allow production of large amounts of cultured plant and animal cells that are conditioned to live in artificial environments. These cultured cells are used for research and production in many scientific fields, i.e., physical science, engineering, agriculture, medicine, and pharmacology, for purely basic studies on morphological changes of cells, for production of interferon, as a promising therapeutic agent, and for development of a variety of cosmetics.

The use of microorganisms in industry is summarized below. In the food industry, important products of microbial production strains include cheese, yogurt, pickles, miso, shoyu (soy sauce), bread, sake (liquors), beer, wine, microbial proteins, and so on. Genetic engineering of these production strains is expected to greatly improve the quality and quantity of these foods. As mentioned previously, in the 1940s microorganisms introduced into the medical and pharmaceutical industries brought on revolutionary changes. The knowledge of microbial physiology and gene manipulation began to be widely used for research and development of new therapeutic drugs and industrial products.

The methods and techniques used to produce antibiotics brought many innovations to scientific research, in medicine, agriculture, and the chemical industries. Some of the currently used fermentation and genetic engineering techniques for large-scale production of drugs evolved during the era of antibiotic development in the 1950s. Microorganisms were used to produce not only foods and drugs, but also various chemicals, including enzymes, aliphatic organic compounds, and amino acids. In the production of amino acids, in particular, chemical synthesis methods have been almost completely replaced by microbial fermentation due to the rise in oil prices and the introduction of recombinant DNA techniques. Genetic engineering studies are underway to increase the yield of enzymes such as glucose isomerase (to produce fructose), plastic polymers, diagnostic enzymes, restriction enzymes, and ligases used in recombinant DNA techniques. The fermentation method used to produce aliphatic organic compounds is based on recombinant DNA techniques; the synthetic chemical industry in general has encountered difficulties because of the unavailability of once abundant and cheap materials, such as oil. In agriculture, the genetics of nitrogen-fixing, symbiotic, leguminous bacteria are being studied to develop techniques that would enable other plants to fix nitrogen and thus increase crop yields. Table 1 shows the principal microorganisms used in industries and their products (Phaff, 1981).

Maintenance and Preservation of Microorganisms for Industrial Use

It would be useful to establish a simple storage method for microorganisms that would ensure maintenance of their initial metabolic activities. Such techniques may be based on modifications of the method used for actinomycetes as described by Hopwood and Ferguson (1969). Actinomycetes are remarkable because of their ability to form complex spores (Cross, 1970) and to produce other structures of unknown function. Selective survivorship occurs among actinomycetes when they are stored under dry, low-temperature conditions; the analysis and elucidation of this survival process will make meaningful contributions to both basic research and industry (Tsuji, 1966). Although research efforts have resulted in production strains for use in a variety of industries, an important consideration is how to store active strains. In an attempt to increase the survival rate of cultures over long periods of storage, procedures have been devised to suit each group of microorganism.

Production strains have specific activities and functions. Unlike research strains, the storage of production strains requires not only maintaining their viability but also ensuring that their metabolic activities remain unchanged. In industrial applications, higher product yields are more likely to be obtained from mutated strains developed for that purpose, rather than from a wild strain. Genetically altered strains, however, are often unstable with respect to productivity, which underlies the difficulty in their storage. At present, no single storage method is applicable to all strains used for industrial production. Numerous methods are tested for each strain and the data are reviewed to find the optimum method for each. The actual storage condition can affect many factors, especially product yield. Because many products vary in type and in production requirements, storage methods vary for production strains. When evaluating storage methods, one must also develop methods to determine activities and to rejuvenate cells. A decrease in cellular activity does not occur uniformly in all cells and, if individual cells are separated and tested for activity, many cells prove to be intact. Cell survival rate and the maintenance of activity do not always correlate (Maruyama, 1982). In general, strains should be stored under

TABLE 1
Main Industrial Organisms and Their Products[a]

Organism	Type	Product
Foods and beverages		
Saccharomyces cerevisiae	Yeast	Baker's yeast, wine, ale, sake
Saccharomycers carlsbergensis	Yeast	Lager beer
Saccharomyces rouxii	Yeast	Soy sauce
Candida milleri	Yeast	Sour French bread
Lactobacillus sanfrancisco	Bacterium	Sour French Bread
Steptococcus thermophilus	Bacterium	Yogurt
Lactobacillus bulgaricus	Bacterium	Yogurt
Propionibacterium shermanii	Bacterium	Swiss cheeses
Gluconobacter suboxidans	Bacterium	Vinegar
Penicillium roquefortii	Mold	Blue-veined cheeses
Penicillium camembertii	Mold	Camembert and Brie cheeses
Aspergillus oryzae	Mold	Sake (rice-starch hydrolysis)
Rhizopus	Mold	Tempeh
Mucor	Mold	Sufu (soybean curd)
Monascus purpurea	Mold	Ang-kak (red rice)
Industrial chemicals		
Saccharomyces cerevisial	Yeast	Ethanol (from glucose)
Kluyveromyces fragilis	Yeast	Ethanol (from lactose)
Clostridium acetobutylicum	Bacterium	Acetone and butanol
Aspergillus niger	Mold	Citric acid
Xanthomonas campestris	Bacterium	Polysaccharides
Amino acids and flavor-enhancing nucleotides		
Corynebacterium glutamicum	Bacterium	L-Lysine
Corynebacterium glutamicum	Bacterium	5′-Inosinic acid and 5′-guanylic acid
Single-cell proteins		
Candida utilis	Yeast	Microbial protein from paper-pulp waste
Saccharomyces lipolytica	Yeast	Microbial protein from petroleum alkanes
Methylophilus methylotrophus	Bacterium	Microbial protein from growth on methane or methanol
Vitamins		
Eremothecium ashbyi	Yeast	Riboflavin

(Continues)

TABLE 1 (*Continued*)

Organism	Type	Product
Pseudomonas denitrificans	Bacterium	Vitamin B_{12}
Propionibacterium	Bacterium	Vitamin B_{12}
Enzymes		
Aspergillus oryzae	Mold	Amylases
Aspergillus niger	Mold	Glucamylase
Trichoderma reesii	Mold	Cellulase
Saccharomyces cerevisiae	Yeast	Invertase
Kluyveromyces fragilis	Yeast	Lactase
Saccharomyces lipolytica	Yeast	Lipase
Aspergillus	Mold	Pectinase and proteases
Bacillus	Bacterium	Proteases
Endothia parasitica	Mold	Microbial rennet
Polysaccharides		
Leuconostoc mesenteroides	Bacterium	Dextran
Xanthomonas campestris	Bacterium	Xanthan gum
Pharmaceuticals		
Penicillium chrysogenum	Mold	Penicillins
Cephalosporium acremonium	Mold	Cephalosporins
Streptomyces	Bacterium	Amphotericin B
		Kanamycins, neomycins
		Streptomycins, tetracyclines, and others
Bacillus brevis	Bacterim	Gramicidin S
Bacillus subtilis	Bacterium	Bacitracin
Bacillus polymyxa	Bacterium	Polymyxin B
Rhizopus nigricans	Mold	Steroid transformation
Arthrobacter simplex	Bacterium	Steroid transformation
Mycobacterium	Bacterium	Steroid transformation
Hybridomas	—	Immunoglobulins and monoclonal antibodies
Mammalian cell lines	—	Interferon
Escherichia coli (via recombinant-DNA technology)	Bacterium	Insulin, human growth hormone, somatostatin, interferon
Carotenoids		
Blakeslea trispora	Mold	Beta-carotene
Phaffia rhodozyma	Yeast	Astaxanthin
Entomopathogenic bacteria		
Bacillus thuringiensis	Bacterium	Bioinsecticides
Bacillus popilliae	Bacterium	Bioinsecticides

[a] Source: H. J. Phaff, *Scientific American* **245**, 52–65 (1981).

conditions that favor maintenance of activities and survival rates. There are only a few reports in the literature that describe how to better maintain activities of useful microorganisms while they are in long-term storage: the scarcity of such reports is due to the fact that research occurring in private corporations and the resulting data are retained as trade secrets.

Improving Storage Techniques

One of the primary goals in technical investigations is to reduce labor costs of storing industrially useful strains.

Shackell (1909) introduced the use of serum and milk for lyophilized specimens; the success of this method served as a significant stimulus to subsequent research. In the 1910s, the serum and skim milk were used as suspending medium and their colloidal protective effects attracted attention. Rogers (1914) reported that the use of skim milk gave beneficial effects on lactic acid bacteria stored by the lyophilization method. In the 1940s, Heller (1941) reported that the addition of metabolic crystal compounds (sucrose, glucose, etc.) with high solubilities enhanced the protective effects of hydrophilic colloids and significantly reduced cell death. Since then, saccharides and other compounds have been tentatively added to the list of protective compounds; Fry and Greaves (1951) reported an increased cell survival rate after addition of 5–10% glucose to the serum bouillon. Many substances have been reported to provide protection against freezing; among the commonly used cryoprotective agents are glycerol (Polge *et al.,* 1949), glutamic acid (Miller and Goodner, 1953), dimethyl sulfoxide (Lovelock and Bishop, 1959), glucose, sucrose, and lactose. It is now a well-established theory that these additives are efficacious in controlling the freezing rate of cells and that the more they are permeable to cell membranes, the greater their protective effects.

The term "L-drying" originated from the expression "drying from the liquid state" as used by its advocate Annear (1958). This method enabled one to reduce significantly the time for drying and thus to economize on labor costs when dealing with a number of strains at the same time. The method, investigated and put into practice by Iijima and Sakane (1970), is widely applied to bacteria, actinomycetes, yeasts, bacteriophages, and some fungi.

Maruyama (1983) compared liquid nitrogen storage, freeze-drying storage, ans freeze-storage at $-20°C$ of slant cultures (Figure 1). Hamada (1982) sealed cotton-plugged slant cultures of actinomycetes with paraffin and froze them at $-20°C$.

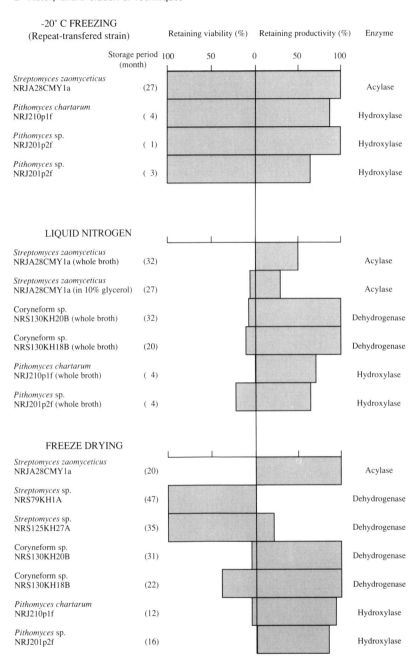

Figure 1. Viability and productivity of enzyme producers maintained by different storage methods. Reprinted with permission from Maruyama, 1983.

A Prospect for the Future

Cryobiology is applicable in scientific research to the maintenance and preservation of microorganisms. In order to conduct highly specialized and/or varied research on the maintenance of cell functions and viability, communication and cooperation are needed among investigators with different expertise. It would be in the best interest of both industrialists and academicians to promote global scientific collaborations. This was a goal, for example, of the International *Streptomyces* Project (ISP) initiated by Shirling and Gottlieb in 1966. In this project, 458 strains of *Streptomyces* and closely related genera were collected and studied in a united effort by investigators worldwide. These strains are now distributed to and stored at institutes, including the American Type Culture Collection (ATCC), the USSR Research Institute for Antibiotics (RIA), Centraal Bureau voor Schimmelcultures (CBS), and the IFO. Almost all of the strains produce antibiotics. The Society for Actinomycetes Japan started the ISP Check Committee, which, since 1969, has checked dried specimens of the ISP strains stored at IFO for their microbiological properties. This quality control check by external experts is the first example of this protocol in the world and has attracted much international interest. However, the ISP does not have the time, money, or resources to evaluate metabolic functions or end products of interest. Public culture collections should be encouraged and supported to perform such tests.

With recent progress in biotechnology, public culture collections have become responsible for maintaining taxonomic characteristics, productivity, resistance or sensitivity to drugs, and susceptibility to mutagens of various cultured organisms, including standard strains. Moreover, their role has been extended to support scientific research. Training of competent research personnel and functioning as leading centers for culture collections necessitate consolidation of finances and a better understanding of the needs of the general public. It is possible that an international cooperation system among public institutes for culture storage could be established under the direction of World Federation for Culture Collections (WFCC).

References

Annear, D. I. (1958). *Aust. J. Exp. Biol.* **36,** 211–222.
Cagniard-Latour, C. (1838). *Ann. Chim. Phys.* **68,** 206–222.
Cross, T. (1970). *J. Appl. Bacteriol.* 33, 95–102.
Fennell, D. I., Raper, K. B., and Flickinger, M. H. (1950). *Mycologia* **42,** 135–147.
Fry, R. M., and Greaves, R. I. N. (1951). *J. Hyg.* **49,** 220–246.
Hamada, M. (1982). *Jpn. J. Freezing Drying* **28,** 63–67.

Hammer, B. W. (1911). *J. Med. Res.* **24**, 527–530.

Hansen, A., and Lund, A. (1939). "Microorganisms and Fermentation," 6th ed., pp. 115, 131. Griffin, London.

Hansen, H. N. (1942). *Phytopathology* **32**, 639–640.

Hansen, H. N., and Smith, R. E. (1932). *Phytopathology* **22**, 953–964.

Hartsell, S. E. (1956). *Appl. Microbiol.* **4**, 350–355.

Heller, G. (1941). *J. Bacteriol.* **41**, 109–126.

Hopwood, D. A., and Ferguson, H. M. (1969). *J. Appl. Bacteriol.* **32**, 434–436.

Hunt, G. A., Gourevitch, A., and Lein, J. (1958). *J. Bacteriol.* **76**, 453–454.

Hwang, S. W. (1960). *Mycologia* **52**, 527–529.

Iijima, T., and Sakane, K. (1970). *Jpn. J. Freezing Drying* **16**, 87–91.

Jensen, H. L. (1961). *Nature (London)* **192**, 682–683.

Kocur, M. (1990). *In* "100 Years of Culture Collections" (L. I. Sly, T. Iijima, and B. Kirsop, eds.), pp. 4–12. Institute for Fermentation, Osaka, Osaka.

Loeffler, L. (1881). *Mitt. Kaiserl. Gesundheitsamte* **1**, 134.

Lovelock, J. E., and Bishop, M. W. H. (1959). *Nature (London)* **183**, 1394–1395.

Lumiere, A., and Cherrotier, J. (1914). *C. Rend. Hebd. Seances Acad. Sci.* **158**, 1820–1821.

Maruyama, H. (1982). *Jpn. J. Freezing Drying* **28**, 58–62.

Maruyama, H. (1983). *Jpn. J. Freezing Drying* **29**, 9–11.

Meyen, J. (1838). *Wigmann Arch. Naturgesch.* **4**, 100.

Miller, R., Jr., and Goodner, K. (1953). *Yale J. Biol. Med.* **25**, 262.

Morton, H. E., and Pulski, E. T. (1983). *J. Bacteriol.* **35**, 163–183.

Pasteur, L. (1860). *Ann. Chim. Phys. [3]* **58**, 323–426.

Phaff, H. J. (1981). *Sci. Am.* **245**, 52–65.

Polge, C., Smith, A. U., and Parkes, A. S. (1949. *Nature (London)* **164**, 666.

Rogers, L. A. (1914). *J. Infect. Dis.* **14**, 100–123.

Schwann, T. (1837). *Ann. Phys. Chem.* **41**, 184–193.

Shackell, L. F. (1909). *Am. J. Physiol.* **24**, 325–340.

Shirling, E. B., and Gotllieb, D. (1966). *Int. J. Syst. Bacteriol.* **16**, 313–340.

Smith, D. (1983). *Trans. Br. Mycol. Soc.* **80**, 333–337.

Tsuji, K. (1966). *Appl. Microbiol.* **14**, 456–461.

van Leewenhoek, A. (1677) *Philos. Trans. R. Soc. London* **11**, 821–831.

Will, H. (1909). *Zentralbl. Bakteriol., Parasitenkd. Infektionskr. Abt. 1* **24**, 405.

CHAPTER 3

Algae

Robert A. Andersen

Background

Introduction

The "algae" comprise an assemblage of plantlike organisms that are found in one prokaryotic and several eukaryotic evolutionary lineages. Physiologically, oxygen-evolving photosynthesis unifies the algal lineages, separating them from protozoa or fungi. Taxonomically, algae are in an artificial group that defies precise definition; like other artificial groups, algae are defined by the absence of characters rather than by their presence. For example, algae can be considered "plants" that *lack* roots, stems, leaves, and embryos (Bold, 1973). Conversely, the individual monophyletic algal lineages *possess* distinctive photosynthetic pigments, storage products, and ultrastructural features (Bold and Wynne, 1985; Lee, 1989).

Historically, the algae have always included prokaryotic and eukaryotic forms (Smith, 1950), but recent studies have suggested that prokaryotic algae are better classified as bacteria (i.e., cyanobacteria) (Stanier *et al.*, 1971). Despite the shared prokaryotic nature of blue-green algae and bacteria, blue-green algae continue to be considered algae by many others (e.g., Bold and Wynne, 1985; Lee, 1989). Less than 10 species of prokaryotic algae have been described using the International Code of Nomenclature of Bacteria; all other species have been described following the International Code of Botanical Nomenclature (Demoulin, personal communication; Komarek, personal communication). In colleges and universities, the prokaryotic algae are taught in phycology courses, not in bacteriology courses. And relevant to this chapter, prokaryotic algae are maintained primarily in algal culture collections, not in bacterial culture collections (e.g., compare catalogs of strains; Gherna *et al.*, 1989; Starr and Zeikus, 1993; Andersen *et al.*, 1991).

Classification and Diversity

Algal systematics is an area of science that is undergoing active research, and like other dynamic scientific fields, new information and new ideas lead to change. The classification of the algae has undergone considerable change during the past 30 years, and it is likely that changes will continue. For example, it once appeared that there were two lineages of prokaryotic algae (Lewin, 1977), but recent evidence suggests there may be only a single lineage (Palenik and Haselkorn, 1992). The eukaryotic algae are now believed to occur in several separate lineages that are separated by protozoan, animal, and fungal lineages (Perasso *et al.*, 1989; Douglas *et al.*, 1991) (Figure 1). The major eukaryotic lineages (i.e., brown, green, and red algal lineages) were firmly established over a century ago (Harvey, 1836, 1846–1851; Rabenhorst, 1864, 1865, 1868), but relationships within and among these lineages are still debated today. The classification presented here (Table 1) represents a moderate approach.

The loss of biodiversity on the planet is a serious problem, and the loss of algal species is as significant as the loss of animal and plant species. The loss of algal species diversity is not documented as thoroughly as for plants and animals, however, because the diversity is not well described, and the geographic range of algae, except for seaweeds, is not well understood. Many "new" species remain to be described, especially in tropical regions, in third world countries, and in oceanic areas, and habitat alteration in these regions may be an unrecognized source leading to the loss of algal biodiversity.

Some algal lineages (e.g., blue-green or red algae) are ancient lineages with a well-substantiated geological history, and apparently other algae evolved relatively recently (e.g., certain classes of chromophyte algae) (Figure 2) (Tappan, 1980). For example, the diatoms appear in the fossil record about 100 million years ago (Tappan, 1980), or about the same time that flowering plants first appear. By comparison, mammals first appear in the fossil record over 200 million years ago. Compared to diatoms or red algae, the fossil record for some other algal groups is poor, but it remains apparent that new eukaryotic algal groups have continued to appear from the Precambrian throughout the Paleozoic, Mesozoic, and perhaps even Cenozoic eras (Figure 2) (Tappan, 1980).

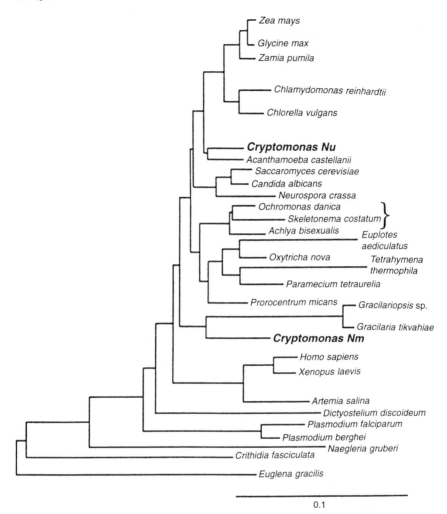

Figure 1. A phylogenetic analysis of 18S rRNA gene sequences (modified from Douglas *et al.*, 1991). Note the six eukaryotic algal lineages interspersed among the animals, fungi, plants, and protozoa. Scale bar equals 10 changes per 100 nucleotides.

Economic Importance

The economic importance of the algae is not always recognized (Lembi and Waaland, 1988). Algae are used as food or in food products, as plant fertilizers, in cosmetics, in manufacturing processes, in biomedical research, and in many other products and processes. Also, algae form the base of

TABLE 1
Higher Level Classification of the Algae,
Including Several Representative Genera[a]

Prokaryotic algae
  Division Cyanophyta (including Prochlorophyta):
    Class Cyanophyceae (*Anabaena, Gloeocapsa, Microcystis, Nostoc,
      Oscillatoria, Prochlorococcus, Synechococcus*)
Eukaryotic algae
  Green algae
    Division Charophyta: Class Charophyceae (*Chara, Cosmarium,
      Nitella, Spirogyra, Zygnema*)
    Division Chlorophyta: Class Chlorophyceae (*Chlamydomonas,
      Chlorella, Tetraspora, Volvox*), Class Ulvophyceae (*Bryopsis,
      Codium, Ulva*)
    Division Prasinophyta: Class Prasinophyceae (*Mantoniella,
      Micromonas, Pyramimonas, Tetraselmis*)
  Chromophyte algae
    Division Bacillariophyta: Class Bacillariophyceae (*Achnanthes,
      Navicula, Nitzschia, Pinnularia*), Class Coscinodiscophyceae
      (*Actinocyclus, Coscinodiscus, Ditylum, Odantella, Thalassiosira*),
      Class Fragilariophyceae (*Fragilaria, Synedra*)
    Division Chrysophyta: Class Chrysophyceae (*Chromulina,
      Chrysocapsa, Dinobryon, Ochromonas, Uroglena*), Class
      Synurophyceae (*Chrysodidymus, Mallomonas, Synura*)
    Division Dictyochophyta: Class Dictyochophyceae (*Dictyocha,
      Apedinella, Pseudopedinella*), Class Pelagophyceae (*Pelagococcus,
      Pelagomonas*)
    Division Eustigmatophyta: Class Eustigmatophyceae (*Eustigametos,
      Nannochloropsis*)
    Division Phaeophyta: Class Phaeophyceae (*Ectocarpus, Laminaria,
      Macrocystis, Undaria*)
    Division Prymnesiophyta: Class Prymnesiophyceae
      (*Chrysochromulina, Coccolithus, Emiliania, Prymnesium, Pavlova*)
    Division Raphidophyta: Class Raphidiophyceae (*Chattonella,
      Fibrocapsa, Gonyostomum*)
    Division Xanthophyta: Class Xanthophyceae (*Mischococcus,
      Tribonema, Vaucheria*)
  Red algae
    Division Rhodophyta: Class Rhodophyceae (*Bangia, Eucheuma,
      Gelidium, Gracilaria, Porphyra, Polysiphonia*)

(continues)

TABLE 1 (Continued)

Cryptomonads
    Division Cryptophyta: Class Cryptophyceae (*Chroomonas,*
      Cryptomonas, Hemiselmis, Komma, Rhodomonas)
Dinoflagellates
    Division Pyrrhophyta: Class Pyrrhophyceae (= Dinophyceae)
      (*Alexandrium, Dinophysis, Gonyaulax, Peridinium, Pyrocystis*)
Euglenoids
    Division Euglenophyta: Class Euglenophyceae (*Astasia, Diplonema,*
      Euglena, Phacus)
Glaucophytes
    Division Glaucophyta: Class Glaucophyceae (*Cyanophora,*
      Glaucocystis)

[a] Some classes contain large numbers of species and are well defined by specific characteristics, whereas other classes have only a few species and/or may be poorly characterized. For the estimated number of species in each class, see Andersen (1992).

the food chain for 71% of the world's surface because they are the only significant primary producers in the oceans, i.e., commercial fisheries and aquaculture depend, directly or indirectly, on algae.

Algae, especially via algal products, account for billions of dollars worth of food and industrial products each year. Commercial products are derived from field collections of native algae (especially seaweeds), from cultivated algae raised either in natural habitats or in man-made ponds and laboratories, and from fossilized deposits (e.g., crude oil, diatomite, limestone) (Lembi and Waaland, 1988; George, 1988).

FOOD PRODUCTS

Algae, as processed and unprocessed food, have a commercial value of several billion dollars annually (Abbott, 1988; Druehl, 1988; Jassby, 1988a). Approximately 500 species are used as food or food products for humans, and about 160 species are valuable commercially (Abbott, 1988). People in the western societies use algal products indirectly in commercially prepared foods, especially puddings, pies, ice cream, etc. People in oriental and Pacific island societies consume seaweeds directly in soups, rice dishes, and meat dishes. Hawaiians were eating about 75 species of algae in the 1800s, and during the 1980s more than 50 species were still being consumed (Abbott, 1988).

The red alga *Porphyra* is the most important alga that is eaten directly as food and is commonly used to make sushi. Worldwide, the retail value

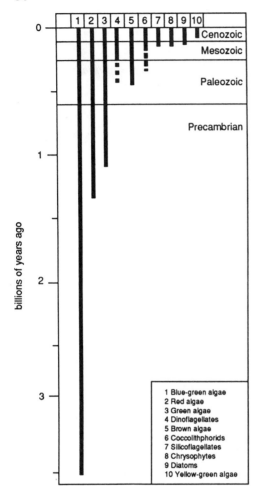

Figure 2. The geological record for algae. Data from Tappan (1980) and Taggart and Parker (1976). Reprinted with permission from Andersen (1992).

of the *Porphyra* harvest for 1986 was about $1 billion (Mumford and Miura, 1988), and this amount is increasing annually. Japan produces about 10 billion sheets of *Porphyra* (nori) each year, or about 3×10^6 kg dry weight; this amounts to about 60% of the world's commercial harvest of *Porphyra*.

The utilization of many seaweeds is restricted biogeographically. *Porphyra*, on the other hand, is a widespread genus with many species, and it has been utilized for food by many cultures, including the oriental cultures in Asia, the Native Americans (American Indians) in the Western Hemisphere, the island cultures in the Pacific (including the Maoris of New Zealand), and the Europeans in Ireland and Britain (Mumford and Miura, 1988). Originally, fishermen in Japan placed tree branches or bamboo shoots in the bays, and (unbeknown to the fishermen) spores adhered to the branches and eventually gave rise to *Porphyra* sporophyte plants. Drew (1949) described the life cycle for *Porphyra*. Quickly thereafter, the nori industry expanded because the Conchocelis stage was grown indoors in oyster shells and then the oyster shells were transported to net farms in the bays (Miura, 1975; Tseng, 1981). Today, *Porphyra* is raised in many regions of the world. *Porphyra* traits follow mendelian genetics and have been relatively easy to select (Miura, 1979; van der Meer, 1988; Patwary and van der Meer, 1992). The rapid life cycle, combined with selection at both the sporophyte stage and the conchocelis stage, has led to rapid development of strains genetically superior for commercial purposes. Currently, hundreds of varieties exist for *Porphyra tenera* and *Porphyra yezoensis*. Most varieties are protected by law, and royalties are paid to the originator (Mumford and Miura, 1988). The use of algae for food is not limited to humans, however. Phytoplankton are used for feeding a variety of animals, especially shrimp and shellfish.

"Farming in the sea" is termed mariculture. The predominant form of algal mariculture is the farming of seaweeds in harbors, bays, and estuaries (Ohno and Critchley, 1993). Interestingly, agriculturally important plants and animals were domesticated before recorded history, and since the beginning of recorded history about 7000 years ago, no major, economically important agricultural organisms have been domesticated. Conversely, mariculture has developed in the past few centuries. For example, *Porphyra* has been used as food by humans for at least 1500 years (Tseng, 1981), but *Porphyra* farming began only 300 years ago (Okazaki, 1971; Tseng, 1981). The domestication of phytoplankton (e.g., *Dunaliella* spp., *Skeletonema costatum, Thalassiosira pseudonana*) for mariculture dates from the 1950s and 1960s. Rapid growth in shellfish (clams, mussels, oysters) and shrimp farming has been possible in part because of advances in techniques for mass culture of phytoplankton. Phytoplankton are fed to larval shellfish in hatcheries prior to the release of "seed" shellfish in natural waters.

INDUSTRIAL PRODUCTS

Algal products are used in the preparation or manufacture of many nonfood products. The agars, carrageenans, and alginates, collectively termed phycocolloids, are a major source of industrially important algal products (Lewis *et al.,* 1988). Agar and agaroses are used in medical and biological sciences for culture media and for gel electrophoresis. Agars are also used in many other products, including ion-exchange and affinity chromatography, pharmaceutical products, and fruit fly foods (Armisen and Galatas, 1987). Carrageenans are used as binders and thickeners in a wide variety of pastes, lotions, and water-based paints (Stanley, 1987). Alginates are used to bind textile printing dyes, to stabilize paper products during production, to coat the surfaces of welding rods, to serve as binders and thickeners in numerous pharmaceutical products, and to act as binders in animal feed products (McHugh, 1987). Diatomite (fossilized diatom frustules) is used as a blocking agent in the manufacture of plastics, as a mild abrasive in toothpastes and household cleaners, as a filtration compound for a wide variety of liquids, and as a binder for paints and pastes.

Medical Importance

Algae are rarely used directly as cures for diseases, but there are some examples of antibiotic activity, vermifuge activity, antitumor activity, and goiter treatment (see Stein and Borden, 1984; Cooper *et al.*, 1983). However, algal extracts are important in the manufacture of many pharmaceutical products, acting as emulsifiers and binding agents for syrups, tablets, capsules, and ointments (see above).

Algae cause few human diseases. Perhaps the best known algal disease is protothecosis, a subcutaneous infection that creates lesions. Protothecosis results from infections in humans of the achlorotic green alga *Prototheca* (Schwimmer and Schwimmer, 1955; Stein and Borden, 1984), and a similar disease is caused by a chlorophyll-containing species (Jones *et al.*, 1983). [For other public health aspects, see Jassby (1988b)].

Algal toxins are the most significant threat to human health, and some toxins can cause severe illness and death. Toxic algae in marine environments cause problems for humans because shellfish may feed on toxic phytoplankton, and in turn the toxified shellfish are consumed by humans (Shumway, 1990). Toxic algae in freshwater ponds and lakes are a serious threat to livestock because livestock drink the toxic water (Falconer *et al.*, 1983; Jackson *et al.*, 1984; Gorham and Carmichael, 1988). Recent studies suggest that toxic algal blooms may be increasing, possibly as a result of human activity (Hallagraeff, 1993).

To date, algae have not been important sources of natural products used as medical drugs, due primarily to the limited research on algal natural products (Baslow, 1969). In Japan, where *Porphyra* is an economically important alga, some research with medical applications has been completed. A diet of *Porphyra* (and, in some cases, other seaweeds) lowered blood cholesterol levels in rats (Abe and Kaneda, 1972), was effective against stomach ulcers (Sakagami *et al.,* 1982), reduced mammary cancer in mice (Yamamoto and Maruyama, 1984), and lowered intestinal cancer in rats (Yamamoto and Maruyama, 1985). The search for medically important natural products from algae is becoming easier as more and more algal species are brought into culture. The maintenance of algae may be a key factor in exploiting algae as a source of natural products.

Methods for Preservation

Cryopreservation

CRYOPRESERVATION AS A MEANS OF ALGAL MAINTENANCE

The goal of cryopreservation is indefinite storage of a strain at very low temperatures, where genetic stability is maintained. Cryopreservation is widely used for animal cells, bacteria, fungi, plant cells, and protozoa (Ashwood-Smith and Farrant, 1980; Nerad and Daggett, 1992). Cryopreservation is usually accomplished by treating living cells with a cryoprotectant, cooling the cells, and then storing the cells in liquid nitrogen (or its vapors) so that living cells remain indefinitely viable for reculturing (Hwang and Hudock, 1971; Ashwood-Smith and Farant, 1980; Morris, 1981; McLellan, 1989; Nerad and Daggett, 1992). When cells are held continuously at or near the boiling point of nitrogen ($-196°C$), they have no significant biological activity, and if the cells are shielded from ultraviolet light, X-rays, etc., the DNA should be perfectly preserved and therefore genetically stable. For successful cryopreservation, a substantial percentage of frozen cells, when thawed and recultured, must grow and divide normally. If the "surviving" cell population is too small, then it is possible that a subpopulation of "freeze-resistant" cells may flourish, replacing the "normal" cells. "Cryopreservation selection" is a type of evolutionary change, and absolute genetic conservation fails when selection occurs.

Cryopreservation has been used successfully for green algae, red algae, haptophytes, euglenophytes, diatoms, and blue-green algae (Morris, 1978, 1981; McLellan, 1989; Beaty and Parker, 1990, 1992; Lee and Soldo, 1992; Cañavate and Lubian, 1994; Kuwano *et al.,* 1994). Morris (1978) successfully preserved 252 of 284 strains of chlorococcalean green algae, and Beaty and Parker (1992) had 78% success in cryopreserving 365 strains of eukaryotic algae. However, despite this success, the use of cryopreservation in algal maintenance is less common than for other organisms. For example, cryopreservation is not employed at most algal culture collections; the Centre for Culture of Algae and Protozoa (England) is a notable exception (McLellan,

1989). Some reasons for the limited application of this technique are (1) cost of facilities, (2) cost of operations, (3) lack of cryopreservation personnel trained to work with algae, (4) limited cryopreservation research on many algal groups, (5) lack of consistent success in some algal groups, and (6) absence of success in a few algal groups.

CRYOPROTECTANTS

Cryoprotectants are used to limit ice formation. The successful preservation of a large percentage of living cells depends on limiting ice formation or hypertonic stress. Ice typically forms when the cooling rate is too rapid, and hypertonic stress typically occurs when the cooling rate is too slow. Successful cryopreservation may be influenced by the rate of cooling, the rate of thawing, the type of nonionic cryoprotectant, the use of a "cold-hardening" growth period, the growth phase and its effects (e.g., vacuole size and number, lipid types and content), and the completeness of the growth media used following thawing (Morris, 1976a,b, 1978, 1981). In general, the rate of cooling and the best cryoprotectant are determined empirically (Morris, 1981; Lee and Soldo, 1992). There are several different cryoprotectants in use, the success of which varies with respect to specific strains of algae (Morris, 1981; Nerad and Daggett, 1992). For example, glycerol has been used successfully for many algae (Morris, 1981; Lee and Soldo, 1992). Dimethyl sulfoxide (DMSO) may be better than glycerol for many algae (McLellan, 1989; Lee and Soldo, 1992; Kuwano *et al.,* 1994). In particular, marine algae showed excellent cryopreservation using DMSO (McLellan, 1989; Cañavate and Lubian, 1994; Kuwano *et al.,* 1994). However, because even trace amounts of DMSO inhibit photosynthesis, thorough removal is required following thawing. Methanol is as good as or better than glycerol and DMSO for *Euglena* and *Chlorella.* When *Euglena gracilis* was grown at 20°C and exposed to glycerol, DMSO, and methanol, the percent recovery was much greater with methanol (Figure 3) (Morris, 1981). After cooling to −196°C, *Chlorella emersonii* had over 90% recovery using 1.5 *M* methanol with a cooling rate of 10°C/min (Figure 4) (Morris, 1981). Beaty and Parker (1992) found that Betaine is also an excellent cryoprotectant. Additionally, good preservation may be obtained by either spraying or dipping a monolayer of cells into liquid propane, concentrating the cells, and transferring them to liquid nitrogen for perpetual storage.

In addition to the types of cryoprotectants, the concentration of the cryoprotectant affects the percent recovery of living cells from liquid nitrogen temperatures. For example, *C. emersonii* had different maximum recovery rates using different concentrations of DMSO, and, furthermore, the maximum rate for each occurred at different cooling rates (Figure 5) (Morris, 1981).

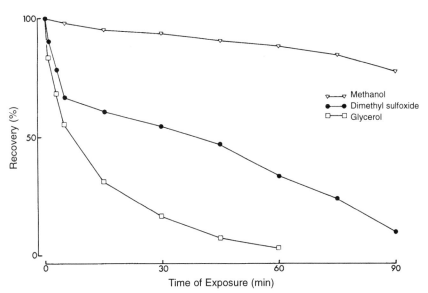

Figure 3. Percent recovery of *Euglena gracilis* after exposure (final concentration = 1.5 *M*) to methanol, dimethyl sulfoxide, or glycerol for different times at 20°C. Reprinted with permission from Morris (1981).

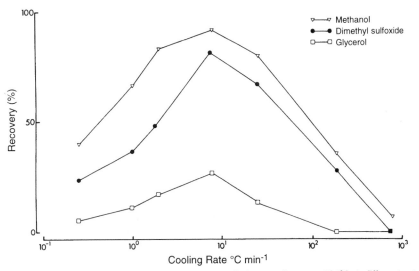

Figure 4. Percent recovery of *Chlorella emersonii* after cooling to −196°C at different rates using methanol, dimethyl sulfoxide, or glycerol (final concentration = 1.5 *M*). Reprinted with permission from Morris (1981).

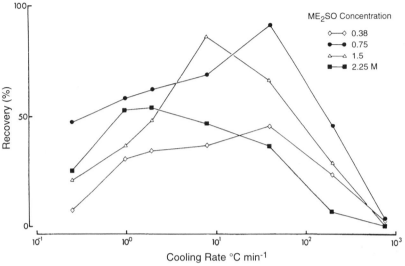

Figure 5. Percent recovery of *Chlorella emersonii* after cooling to −196°C at different rates using four different concentrations of dimethyl sulfoxide. Reprinted with permission from Morris (1981).

CULTURE AGE

Perhaps counterintuitively, the best cryopreservation results are achieved using cells from old, stationary phase cultures rather than using cells from log phase cultures. Cells in old cultures usually lack vacuoles, and vacuolate cells are more difficult to cryopreserve. For example, vacuolate species such as *Chlorella vulgaris* had low (<0.1–10%) recovery rates whereas nonvacuolate strains of *Chlorella protothecoides* had recovery rates greater than 60% (Morris, 1981). Also, as cultures age, the lipid content of the cells changes, and this may provide for more favorable cryopreservation (Morris, 1981).

COLD HARDENING

The percent recovery of cryopreserved cells can also be improved in some algae by employing a "cold hardening" growth period prior to freezing (Morris, 1976b). This is similar to plant frost hardness, which occurs as atmospheric temperatures become cooler during the autumn. Morris (1981) found that *C. emersonii* improved from 0% recovery at its normal 20°C growth temperature to about 45% recovery when it was grown 18 days at 4°C.

RATE OF COOLING

The percent recovery of cryopreserved cells may be affected by the rate of cooling. *Scenedesmus quadricauda* has a 50% recovery rate when cooled at approximately 3°C/min, but a lower recovery when cooled more slowly or more rapidly (Figure 6) (Morris, 1981). Similarly, *Chlamydomonas nivalis* has a bell-shaped distribution of percent recovery, and the maximum recovery (60%) occurs with a cooling rate of 10°C/min (Figure 6) (Morris, 1981).

Freeze-Drying

Freeze-drying techniques have been employed successfully for the preservation of algae (Daily and McGuire, 1954; Holm-Hansen, 1964, 1973; Tsuru, 1973; Corbett and Parker, 1976; McGrath et al., 1978; Nerad and Daggett, 1992). Compared to cryopreservation, the facilities and operation costs are much reduced (Holm-Hansen, 1973). However, freeze-drying is less reliable than cryopreservation. McGrath et al. (1978) point out that freeze-drying, compared to continual subculturing, minimizes the risk of contamination and genetic change. Also, where applicable, lyophilized cultures may be distributed from culture centers with greater reliability than actively growing cultures.

Cells are frozen, usually with an additive, and the water is removed from the frozen cells using a vacuum pumping system (Holm-Hansen, 1973). The dried sample is sealed in an ampoule under vacuum or in a nitrogen gas atmosphere, and thereafter the freeze-dried cells are stored at room

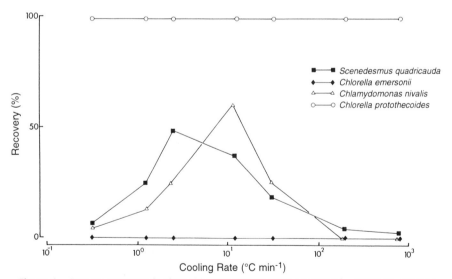

Figure 6. Percent recovery for four different species after cooling to −196°C at different rates. Reprinted with permission from Morris (1981).

temperature. Actively growing cells are obtained by adding culture medium to the freeze-dried cells.

FREEZE-DRYING METHODS

The additives used during the freezing process include milk solids, skim milk, animal sera, and sucrose (Daily and McGuire, 1954; Holm-Hansen, 1964, 1973; Tsuru, 1973; Corbett and Parker, 1976; McGrath et al., 1978). McGrath et al. (1978) recovered viable cells from 39 of 106 strains tested, and of the additives tested, skim milk and sucrose provided the highest recovery percentages, 20 and 12%, respectively. As with cryopreservation, older stationary phase cultures are preferable to log phase cultures. Freezing methods include plunging the cells with additives into liquid nitrogen or acetone–dry ice (Tsuru, 1973).

Once the cells are lyophilized, they are placed in ampoules and sealed. To reduce oxidative degradation, the ampoules are filled with nitrogen gas or sealed under a vacuum. Typically, freeze-dried cells are stored at room temperature (McGrath et al., 1978), but degradation may be slowed if the vials were stored at $-80°C$.

The recovery of freeze-dried cells is typically very low, often under 5% (Tsuru, 1973). Freeze-drying has not been actively pursued during recent years, but the use of slow, controlled freezing methods developed for cryopreservation may provide higher success rates compared with the success rates obtained using "cold plunge" methods. The selection of "freeze-resistant" cells may occur under low recovery conditions. Also, the viability of freeze-dried cells decreases with time, further exacerbating the selection of freeze-resistant cells (Tsuru, 1973; M. M. Watanabe, personal communication).

Serial Transfer

HISTORICAL PERSPECTIVE

Algal culture techniques are described in detail in several books and articles (e.g., Kufferath, 1930; Bold, 1942; Pringsheim, 1946; Lewin, 1959; Stein, 1973; Guillard, 1975; Richmond, 1986). Many of the basic methods and media that are used today were developed in the late 1800s and early 1900s (for review, see Allen and Nelson, 1910; Allen, 1914). The importance of sterile technique and the concept of perpetual maintenance were adapted from microbiological research (Beijerinck, 1890, 1891, 1893; Miquel, 1893a–e); single-cell isolation techniques using a capillary pipette or serial dilutions were developed (Miquel, 1893d,e), axenic algal cultures were established (Miquel, 1893d; Richter, 1903), and culture media were prepared by adding chemical compounds to lake water, seawater, or distilled water (Miquel, 1893b,c; Molisch, 1895, 1896; Benecke, 1898). It was discovered that the exact concentrations and ratios of chemicals were not always critical, and subsequently other scientists have modified these early media recipes. Furthermore, early workers recognized that chemicals contained impurities, implying that unknown contaminating trace nutrients were being added inadvertently to the culture medium, contributing to better algal growth (Allen and Nelson, 1910; Allen, 1914). The situation was summarized succinctly by Provasoli and Pintner (1960), who explained why early media required only Fe and later media required Co, Cu, Mn, Mo, V, and Zn: "It is worth noting that the industrial methods of purification of the 'chemically pure' salts have undergone many changes since Knop's time resulting in an ever-changing sets of impurities." The "impurities" of distilled water, seawater, and glassware were recognized, also, metal toxicity was described, and the importance of pH, vitamins, and iron (as a trace metal) was established (Allen and Nelson, 1910; Allen, 1914).

TYPES OF CULTURE MEDIA

Enrichment media and artificial media are the two major types of culture media. An enrichment medium is prepared either by adding soil or soil extracts to distilled water or to natural water or by adding nutrient chemicals to natural water (e.g., lake water or seawater). An artificial medium uses "pure" water and "pure" chemicals only; it does not include additions of undefined soil or natural lake water or seawater. However, it is very important to recognize that unknown impurities are present in even the most carefully prepared artificial medium (see Freifelder, 1964).

Soil- and water-enriched media are best for maintenance of algal cultures. Good soils contribute inorganic and organic materials, excellent algal growth occurs, and culture-induced morphological changes are reduced. Prát et al. (1972) compared algae grown in Medium #41 with algae grown in Medium #41 plus soil. In almost all cases, growth was enhanced by the addition of soil (Table 2).

Marine media prepared by chemically enriching seawater are common and widely used (McLachlan, 1973); however, the use of lake or river water for chemically enriched cultures is uncommon today. Conversely, artificial media are common and very successful for the growth of many freshwater algae (see Starr and Zeikus, 1993), and artificial media are rarely used for culturing marine algae, due in part to the complexity of seawater. A carefully defined, artificial seawater medium was prepared to minimize or exclude known contaminants for the purpose of studying trace elements (Morel et al., 1979; Price et al., 1989). Although it is impossible to exclude all undesirable substances, in practical situations, the techniques used in preparing this medium seem sufficient for precise experiments involving carefully measured nutrients. Similarly, water quality, glass/plasticware cleanliness,

TABLE 2
Growth of Algal Strains in Medium #41
Plus Soil[a]

Number of strains	Affect on growth
9	0.5–1.0×
40	1.0–1.5×
50	1.5–2.0×
40	2.0–3.0×
6	3.0–4.0×
4	4.0–5.0×
2	5.0–6.0×
2	6.0–7.0×
1	17×

etc., impact culture media in the same manner as chemical quality, i.e., the presence of undesirable or toxic materials and the absence of required substances affect algal growth. Culture media and culturing vessels and utensils can be sterilized by steam autoclaving, pasteurization, filtration, or microwaves.

CULTURE FACILITIES

The needs of a culture facility vary with the user. For some, ordinary laboratory benches illuminated by north-window lighting are sufficient. Others may require commercial culture boxes, walk-in culture rooms, large-volume tanks, or even outdoor ponds. When light and temperature settings are critical, precise regulation of light and temperature is best achieved using digital control timers; mechanical clock timers are too imprecise. In addition to the actual growth-chambers, the culture facility should have areas for transferring cultures aseptically, for microscopic observation, for storage of glassware and media, for preparing and sterilizing media, etc. Often, microalgae are easier to culture than macroalgae because the entire organism is a single cell, epiphytes are less common, and techniques developed for general microbiology can be applied more easily. [For a general review, see Stein (1973).]

Future Needs and Directions

A large majority of algal strains are maintained through perpetual subculturing of living stocks. The use of cryopreservation and other preservation techniques is rare for algal cultures, and clearly, the utilization of these preservation techniques must be expanded dramatically. Maintenance of subcultured living stocks requires extensive investment in labor, in illuminated and refrigerated culture boxes, and in supplies, for example. Furthermore, a routine of subculturing living stocks increases the chances for contamination and mislabeling of strains. And finally, the "evolution" of strains over years of subculturing leads to genetic changes over time. Thus, in some instances, it may not be possible to repeat scientific research that was conducted on a particular strain one or two decades ago. A more widespread use of cryopreservation techniques should allow culture collections to maintain more strains, and to preserve algal biodiversity, maintaining strains even after their habitats are destroyed.

Culture Collections Which Maintain Algal Strains

Australia

Australian Water Quality Centre
Private Mail Bag 3
Salisbury, South Australia 5108
Australia
Collection abbreviation: ACWQ
Curator: Mr. Peter Baker
Telephone: (61)(8) 259-0338
Fax number: (61)(8) 259-0228

Australian Collection of Microorganisms
Centre for Bacterial Diversity and Identification
Department of Microbiology, University of Queensland
Brisbane, Queensland 4072
Australia
Collection abbreviation: ACM
Director: Dr. L. I. Sly
Telephone: (61)(7) 365-4617
Fax number: (61)(7) 365-4620
   Catalog available.

Commonwealth Scientific and Industrial Research Organization
CSIRO Division of Fisheries Research
GPO Box 1538
Hobart, Tasmania 7001
Australia
Collection abbreviation: CSIRO
Director: Dr. Susan Blackburn
Curator: Jeannie-Marie LeRoi
Telephone: (61)(02) 32-5222

Fax number: (61)(02) 32-5000
email: Sue.Blackburn@ML.CSIRO.au
   Catalog available.

Melbourne University Culture Collection
School of Botany, University of Melbourne
Parkville, Victoria 3052
Australia
Collection abbreviation: MUCC
Curator: Dr. David Hill
Telephone: (61)(3) 344-5131
Fax number: (61)(3) 347-5460
email: u6067930@hermes.unimelb.edu.au

Murdoch University Algal Culture Collection
School of Environmental and Life Sciences
Murdoch, Western Australia 6150
Australia
Collection abbreviation: MUR
Director: Dr. Michael A. Borowitzka
Telephone: (61)(9) 360-2333
Fax number: (61)(9) 310-3505
email: borowitz@possum.murdoch.edu.au

The Australian Collection of Marine Microorganisms
Department of Biomedical and Tropical Veterinary Science
James Cook University
Townsville, Queensland 4811
Australia
Collection abbreviation: ACCM
Curator: Dr. Paul Muir
Dr. Warren Shipman
Telephone: (61)(77) 814-278
Fax number: (61)(77) 791-526
   Catalog available.

Bulgaria

Plovdiv Algal Culture Collection
Department of Botany
"Paissi Hilendarsky" University of Plovdiv
Todor Samodumov Street Nr. 2
Plovdiv 4000
Bulgaria
Collection abbreviation: PACC
Director: Prof. Dimitar Vodenicharov
Curator: D. Belkinova
Fax number: 32-238607
   Catalog available.

Canada

North East Pacific Culture Collection
Department of Oceanography
University of British Columbia
6270 University Boulevard
Vancouver, British Columbia V6T IW5
Canada
Collection abbreviation: NEPCC
Director: Dr. F. J. R. Taylor
Curator: Elaine P. Simons
Telephone: 604-228-4378
Fax number: 604-228-6091
    Catalog available.

University of Toronto Culture Collection
Department of Botany
University of Toronto
Toronto, Ontario M5S 3B2
Canada
Collection abbreviation: UTCC
Director: Dr. C. Nalewajko
Curator: J. C. Acreman
Telephone: 416-978-3641
Fax number: 416-978-5878
email: jacreman@botany.utoronto.ca
    Catalog available.

Czech Republic

Culture Collection of Autotrophic Organisms
Institute of Botany
Czechoslavak Academy of Sciences
Dukelska 145,
Trebon CZ-379 82
Czech Republic
Collection abbreviation: CCALA
Curator: Dr. Lukavsky
Telephone: 42-333-2522
Fax number: 42-333-2391
email: hauser@sigma.jh.jcu.cz
    Catalog available.

Denmark

Scandinavian Culture Centre for Algae and Protozoa
Botanical Institute
X. Farimagsgade 2D,
Copenhagen K, DK-1353
Denmark
Collection abbreviation: SCCAP

Curator: Niels Henry Larsen
Telephone: 35-32-23-03
Fax number: 35 32 23 21
email: sccap@bot.ku.dk
Restrictions: closed during August
   Catalog available.

Finland

University of Helsinki Algal Culture Collection
Department of Applied Chemistry and Microbiology
University of Helsinki
P.O. Box 27
Helsinki, FIN-0014
Finland
Collection abbreviation: UHACC
Curator: Dr. Kaarina Sivonen
Telephone: 358-0-708-5288
Fax number: 358-0-708-5212
email: kaarina.sivonen@helsinki.fi

France

Caen Algal Culture Collection
Université de Caen
Laboratoire de Biologie et Biotechnologies Marines (Equipe Phycologie)
Esplanade de la Paix
Caen 14032
France
Collection abbreviation: CAEN
Director: Dr. Chantal Billard
Telephone: (33) 31-45-58-85
Fax number: (33) 31-45-56-00

Centre d'Oceanologie de Marseille
CNRS URA n41
Station Marine d'Endoume
Rue de la Batterie des lions
Marseille 13007
France
Collection abbreviation: COM
Director: Dr. Brigitte Berland
Telephone: (33) 91 04 16 38
Fax number: (33) 91 04 16 35

Centre de Nants
Laboratoire Phycotoxincs et Nuisances
BP 1105
Nantes 44311, Cedex 03
France
Collection abbreviation: CN

Director: Dr. Patrick Lassus
Telephone: (33) 40 37 41 30
Fax number: (33) 40 37 40 73

Pasteur Culture Collection of Cyanobacterial Stains in Axenic Culture
Unite' de Physiologie Microbienne
Institut Pasteur
28 rue du Docteur Roux
Paris 75724, Cedex 15
France
Collection abbreviation: PCC
Director: Rosmarie Rippka
Curator: Michael Herdman
Telephone: (33-1)4568-8416
Fax number: (33-1)4061-3042
email: cyano@pasteur.fr
     Catalog available.

Germany

Sammlung von Algenkulturen
Pflanzenphsiologisches Institut der Universität Gottingen
Nikolausberger Weg 18
Gottingen D-37073
Germany
Collection abbreviation: SAG
Director: Prof. Dr. U. G. Schlösser
Fax number: 0551-397871
     Catalog available.

Sammlung von Conjugaten-Kulturen
Institut für Allgemeine Botanik der Universität Hamburg
Ohnorststrasse 18
Hamburg D-2000
Germany
Collection abbreviation: SVCK
Director: Dr. Monika Engels

Greece

National Centre for Scientific Research
153 10 AG. Paraskevi Attikis
POB 60228
Greece
Curator: Dr. Lydia Ignatiades
Collection abbreviation: NCSR
Telephone: 651-311-19

University of the Aegean Culture Collection
Department of Environmental Studies (UACC)
Kavetsou 12-14

Mytilini 81100
Greece
Director: Prof. M. Karydis

India

National Collection of Industrial Microorganisms
Biochemistry Division (NCIM)
National Chemical Laboratory, CSIR
Poona, Mahasahtra 411 008
India
Director: Mr. S. R. Modak

Israel

Israel Oceanographic and Limnological Research
LTD
P.O. Box 8030
Haifa 31080
Israel
Collection abbreviation: IOLR
Curator: Prof. Ami Ben-Amotz
Telephone: 04-515202
Fax number: 04-511911

Japan

Akashiwo Research Institute of Kagawa Prefacture
Yashima-Higashi-machi, Takamatsu
761-01, Japan
Collection Abbreviation: KAGAWA
Director: Dr. Chitari Ono
Curator: Dr. Sadaaki Yoshimatsu
Telephone: (0878)-43-6511
Fax Number: (0878)-41-8133

Faculty of Agriculture
The University of Tokyo
Yayoi, Bunkyo-ku
Tokyo 113
Japan
Collection abbreviation: FAUT
Curator: Prof. Yasuwo Fukuyo

Institute of Molecular and Cellular Biosciences
The University of Tokyo
1-1-1 Yayoi, Bunkyo-ku
Tokyo 113
Japan
Collection abbreviation: IAM

Director: Prof. Junta Sugiyama
Telephone: 81-3-3812-2111
Fax number: 81-3-3818-0444
    Catalog available.

Kyoto University Culture Collection
Laboratory of Microbiology
Department of Fisheries
Faculty of Agriculture
Kyoto 606
Japan
Collection abbreviation: KUCC
Director: Prof. Yuzaburo Ishida
Telephone: (075) 763-6217
email: h51405@sakura.kudpc.kyoto-u.ac.jp

Marine Biotechnology Institute
Kamaishi Laboratories
3-75-1 Heita
Kamaishi-shi, Iwate 026
Japan
Collection abbreviation: MBI
Director: Dr. Shigetoh Miyachi
Curator: Dr. Hisato Ikemoto
Telephone: 81-193-266538
Fax number: 81-193-266584

Department of Marine Science
School of Marine Sciences and Technology
Tokai University
Orito, Shimizu, Shizuoka 424
3-20-1
Japan
Collection abbreviation: MSTU
Director: Prof. Kaori Ohki

National Institute for Environmental Studies
Microbial Culture Collection
16-2 Onagawa, Tsukuba
Ibaraki 305
Japan
Collection abbreviation: NIES
Director: Dr. Makoto M. Watanabe
Curator: Dr. Nozaki
Telephone: 0298-51-6111
Fax number: 0298-51-4732
    Catalog available.

Mexico

La Coleccion de Microalgas del CICESE
Centro de Investigacion Cientifica Y de Educacion Superior de Ensenada

Km. 107 Carretera
Tijuana, Ensenada
Mexico
Collection abbreviation: CICESE
Curator: M.L. Trujillo Valle
Telephone: 91-617-448-80
Fax number: 91-617-448-80
  Catalog available.

New Zealand

New Zealand Oceanographic Institute Phytoplankton Culture Collection
National Institute of Water and Atmospheric Research
P.O. Box 14-901
Kilbirnie, Wellington
New Zealand
Collection abbreviation: NZOI
Curator: Dr. F. Hoe Chang
Telephone: 64-4-38 60300
Fax number: 64-4-38 62153

The Cawthron Microalgae Culture Collection
Cawthron Institute
Private Bag 2, Nelson
New Zealand
Collection abbreviation: CAWT
Curator: Ms Maggie Atkinson
Telephone: (64)(3) 548-2319
Fax number: (64)(3) 546-9464
  Catalog available.

Norway

Culture Collection of Algae
Norwegian Institute for Water Research
P.O. Box 173 Kjelsås
Blindern N-411, Oslo, Norway
Collection Abbreviation: NIVA
Curator: Dr. Olav M. Skulberg
Fax Number: 47-2235280
Catalog available

University of Oslo
Department of Biology, Marine Botany
P.O. Box 1069
Blindern N-0316, Oslo
Norway
Collection abbreviation: UIO
Director: Dr. Jahn Throndsen
Fax number: 47-22854438

People's Republic of China

Freshwater Algae Collection
Institute of Hydrobiology
The Chinese Academy of Sciences
Wuhan
People's Republic of China
Collection abbreviation: FACHB
Director: Dr. Li-Rong Song
  Catalog available.

Collection of Asian Phytoplankton
Institute of Oceanology Academia Sinica
7 Nanhai Road
Qingdao
People's Republic of China
Collection abbreviation: CAP
Curator: Dr. C. Y. Wu
  Catalog available.

Institute of Hydrobiology Algal Collection
Jinan University
Guangzhou
People's Republic of China
Collection abbreviation: JINAN
Curator: Prof. Yu-zao Qi

The Philippines

Algal Culture Collection of University of Philippines—Los Banos
Museum of Natural History, Institute of Biological Sciences
P.O. Box 169
University of Philippines at Los Banos
Laguna 4031
The Philippines
Collection abbreviation: UPLB
Curator: Dr. Milagrosa R. Martinez-Goss
Telephone: (63-94) 2570
Fax number: (63-94) 3249

Blue-Green Algal Collection at IRRI
International Rice Research Institute
Soil Microbiology Division
P.O. Box 933
Manila
The Philippines
Collection abbreviation: IRRI
Curator: Dr. J. K. Ladha
Fax Number: (63-2) 817-8470
email: j.k.ladha@cgnet.com

Marine Science Institute
University of Philippines

U.P.P.O. Box 1, Diliman
Quezon City 1101
The Philippines
Collection abbreviation: MSIUP
Director: Dr. Edgardo D. Gomez
Curator: Rhodora Corrales
Telephone: 98-224-71 (7414)
Fax number: 632-921-5967
email:rhod@msi.upd.edu.ph

Russia

Algological Collection
Institute of Plant Physiology RAS
ul. Botanicheskaya
35 Moscow 127276
Russia
Collection abbreviation: IPPAS
Curator: Dr. E. S. Kuptsova
Telephone: 095-482-4491
095-482-5447
Fax number: 095-482-1685
   Catalog available.

Biological Institute of St. Petersburg State University
Oranienbaumskoye sch. 2
Stary Peterhof
St. Petersburg 198904
Russia
Collection abbreviation: BISPSU
Curator: Dr. Serguei Karpov
Telephone: (812)427-9669
Fax number: (812)428-6649 & 218-1346
email: igor@hg.bio.lgu.spb.su

Collection of Green Algal Cultures
Laboratory of Algology
Botanical Institute by V.L. Komarov of RAS
ul. Prof. Popov 2
St. Petersburg 197376
Russia
Collection abbreviation: LABIK
Curator: Dr. V. M. Andreyeva & A.F. Luknitskaya
Fax number: (812)234-45-12
email: cinran@glas.apc.org.
   Catalog available.

Saudi Arabia

King Fahd University of Petroleum & Minerals
Research Institute

Marine Group, Algal Collection
Dhahran 31261
Saudi Arabia
Collection abbreviation: KFUPM
Curator: Dr. Assad A. Al-Thukair
Telephone: (966)-3-8604180
Fax Number: (996)-3-8604180

South Africa

University of Witwatersrand Culture Collection
Department of Botany
University of Witwatersrand
Private Bag 3 WITS
Johannesburg 2050
South Africa
Collection abbreviation: WITS
Curator: Prof. Richard N. Pienaar
Telephone: (27)(11)-716-2251
Fax number: (27)(11)-403-1429

Spain

Instituto Espanol de Oceanografia
CaboEstay-Canido Ap. de Correos 1552
36280 Vigo, Spain
Collection Abbreviation: VIGO
Director: Dr. I. Bravo
Curator: I. Ramilo
Telephone: 34-86-49-21-11
Fax Number: 34-86-49-23-51
email: insovigo@cesga.es

Sweden

AVD for Marine Ekologi
Department of Marine Ecology
Lund University
Lund
Sweden
Collection abbreviation: AVD
Director: Dr. Lars Gisselson
Telephone: 46-46-108366
Fax number: 46-46-104003

Taiwan

Tungkang Marine Laboratory
Taiwan Fisheries Institute
Tungkang, Pingtung 928

Taiwan R.O.C.
Collection abbreviation: TML
Director: Dr. Mao-Sen Su
Curator: Dr. Huei-Meei Su
Telephone: 886-8-8324121
Fax number: 886-8-8320234

Thailand

Thailand Institute of Scientific and Technological Research
196 Phahonyothin Road
Bangkok 10900
Thailand
Collection abbreviation: TISTR
Curator: Pongtep Antarikanonda

Ukraine

Collection of Algal Strains
Division of Spore Plantae
Institute of Botany by Kholodny of UkrAS
ul. Tereschenkovskaya, 2
Kiev 252601
Ukraine
Collection abbreviation: IBASU-A
Director: Dr. K. M. Sitnick
Curator: Dr. V. P. Yunger
Telephone: 224-51-57
Fax number: 224-10-64
    Catalog available.

United Kingdom

Culture Centre of Algae and Protozoa
Freshwater Biological Institute
The Ferry House, Ambleside
Cumbria LA22 OLP
England
Collection abbreviation: CCAP at IFE
Curator: Dr. J. G. Day
Telephone: 53-9442468
Fax number: 53-9446914
    Catalog available.

Culture Collection of Algae and Protozoa
Dunstaffnage Marine Research Laboratory
P.O. Box 3
Oban, Argyll PA34 4AD
Scotland
Collection abbreviation: CCAP at DML

Curator: Michael F. Turner
Telephone: (01631)562244
Fax number: (01631)565518
email: CCAPN@DML.AC.UK
  Catalog available

Plymouth Culture Collection
Plymouth Marine Laboratory
Citadel Hill
Plymouth, Devon LP1 2PB
England
Collection abbreviation: PLY
Director: Dr. J. Green
Telephone: (44) 752-222772
Fax number: (44) 752-226865

Swansea Algal Research Unit
University of Wales, Swansea
Singleton Park
Swansea, Wales SA2 8PP
United Kingdom
Collection abbreviation: SARU
Director: Dr. Kevin Flynn
Telephone: (44) 792-295726
Fax number: (44) 792-295447

University of Westminster Algal Collection
115 New Cavendish St.
London WIM SJS
England
Collection abbreviation: UW
Curator: Dr. J. Lewis
Telephone: 71-911-5000
Fax: 71-911-5087

United States of America

Algal Collection at Monterey Bay Aquarium Research Institute
Monterey Bay Aquarium Research Institute
Pacific Grove, California 93950
Collection abbreviation: MBARI
Curator: Dr. Chris Scholin
Fax number: 408-647-3779
email: scholin@mbari.org

American Type Culture Collection
Protistology Department
12301 Parklawn Drive
Rockville, Maryland 20852
Collection abbreviation: ATCC
Curator: Dr. Thomas Nerad

Telephone: 800-638-6597
Fax number: 301-816-4361
email: sales@atcc.org
  Catalog available

American Type Culture Collection
Bacteriology Department
12301 Parklawn Drive
Rockville, Maryland 20852
Collection abbreviation: ATTC
Curator: Dr. Robert Gherna
Telephone: 800-638-6597
Fax number: 301-816-4361
email: sales@atcc.org
  Catalog available.

Carolina Biological Supply Company Algal Collection
Algae Department
2700 York Road
Burlington, North Carolina 27215
Collection abbreviation: CBSC
Curator: Daniel James
Telephone: 910-584-0381
Fax number: 910-584-3399
  Catalog available.

Chlamydomonas Genetics Center
Department of Botany, Duke University
DCMB Box 91000
Durham, North Carolina 27708-1000
Collection abbreviation: CGC
Director: Dr. Elizabeth H. Harris
Telephone: 919-613-8164
  Catalog available.

Culture Collection of Algae at University of Texas
Department of Botany
University of Texas at Austin
Austin, Texas 78713-7640
Collection abbreviation: UTEX
Director: Dr. Richard Starr
Curator: Dr. Jeff Zeikus
Telephone: 512-471-4019
Fax number: 512-471-3878
  Catalog available.

Marine Microalgae Research Culture Collection
Florida Department of Environmental Protection
Marine Research Institute
100 Eighth Ave. S.E.
St. Petersburg, Florida 33701

Collection abbreviation: FLORIDA
Curators: Dr. Karen Steidinger and Dr. Carmelo Tomas
Telephone: 813-896-8626
Fax number: 813-823-0166

Freshwater Diatom Culture Collection
Department of Biology, Loras College
Dubuque, Iowa 52001-0178
Collection abbreviation: FDCC
Curator: Dr. David B. Czarnecki
Telephone: 319-588-7231
email: czdiatom@LCAC1.LORAS.edu
  Catalog available.

Milford Laboratory Culture Collection
National Oceanic and Atmospheric Administration
National Marine Fisheries Service
Milford, Connecticut 06460-6499
Collection abbreviation: MIL
Director: Gary H. Wikfors
Curator: Jennifer H. Alix
Telephone: 203-783-4225
Fax number: 203-783-4217

Provasoli-Guillard National Center for Culture of Marine Phytoplankton
Bigelow Laboratory for Ocean Sciences
McKnown Point
West Boothbay Harbor, Maine 04575
Collection abbreviation: CCMP
Director: Dr. Robert A. Andersen
Curator: Dr. Steven Brett
Telephone: 207-633-9630
Fax number: 207-633-9641
email: ccmp@ccmp.bigelow.org
  Catalog available.

University of California-Santa Barbara Culture Collection
Department of Biological Sciences
University of California-Santa Barbara
Santa Barbara, California 93105
Collection abbreviation: UCSB
Curator: Dr. Robert K. Trench
Fax number: 805-893-4724

University of Miami Algal Culture Collection
School of Oceanography, University of Miami
Miami, Florida 33149
Collection abbreviation: MIAMI
Curator: Dr. Larry Brand
Telephone: 305-361-4138

University of Rhode Island
Department of Pharmacognosy & Environmental Health Sciences
College of Pharmacy
University of Rhode Island
Kingston, Rhode Island 02881
Collection abbreviation: URI-1
Director: Dr. Yuzuru Shimizu
Telephone: 401-792-2751
Fax number: 401-792-2181

University of Rhode Island Culture Collection
Graduate School of Oceanography
University of Rhode Island
Narragansett, Rhode Island 02882-1197
Collection abbreviation: URI-2
Curator: Dr. Paul Hargraves
Telephone: 401-792-6241
Fax number: 401-792-6240
email: pharg@gsosun1.gso.uri.edu

Woods Hole Oceanographic Institution
Biology Department, Redfield 3-32
Woods Hole, Massachusetts 02543
Collection abbreviation: WHOI-1
Director: Dr. Donald Anderson
Curator: David Kulis
Telephone: 508-457-2000 ext. 2351
Fax number: 508-457-2169
email: danderson@whoi.edu

Woods Hole Oceanographic Institution
Woods Hole, Massachusetts 02543
Collection abbreviation: WHOI-2
Curator: Dr. John Waterbury
Telephone: 508-457-2000 ext. 2742
Fax number: 508-457-2169
email: jwaterbury@whoi.edu
    Restrictions: cultures sent by overnight courier only.

Vietnam

National Scientific Research Centre of Vietnam
Institute of Biology
Tu Liem, Ha Noi
Vietnam
Collection abbreviation: Vietnam
Curator: Prof. Nguen Huu Thuoc

References

Abbott, I. A. (1988). Food and food products from seaweeds. *In* "Algae and Human Affairs"
    (C. A. Lembi and J. R. Waaland, eds.), pp. 135–147. Cambridge Univ. Press, New York.

Abe, S., and Kaneda, T. (1972). The effect of edible seaweeds on cholesterol metabolism in rats. *Proc. Int. Seaweed Symp.* **9**, 562–565.

Allen, E. J. (1914). On the culture of the plankton diatom *Thalassiosira gravida* Cleve in artificial seawater. *J. Mar. Biol. Assoc. U.K.* **10**, 417–439.

Allen, E. J., and Nelson, E. W. (1910). On the artificial culture of marine plankton organisms. *J. Mar. Biol. Assoc. U.K.* **8**, 421–474.

Andersen, R. A. (1992). Diversity of eukaryotic algae. *Biodiversity Conserv.* **1**, 267–292.

Andersen, R. A., Jacobson, D. M., and Sexton, J. P. (1991). "Provasoli-Guillard Center for Culture of Marine Phytoplankton Catalog of Strains." CCMP, West Boothbay Harbor, ME.

Armisen, R, and Galatas, F. (1987). Production, properties and uses of agar. *In* "Production and Utilization of Products from Commercial Seaweeds," pp. 1–49. FAO/UN, Rome.

Ashwood-Smith, M. J., and Farrant, J., eds. (1980). "Principles and Practice of Low Temperature Preservation in Medicine and Biology." Tunbridge Wells, Pitman Medical Publishing, London.

Baslow, M. H. (1969). "Marine Pharmacology." Williams & Wilkins, Baltimore.

Beaty, M. H., and Parker, B. C. (1990). Investigations of cryo-preservation and storage of eukaryotic algae and protozoa. *J. Phycol.* **26**(2), Suppl., 5.

Beaty, M. H., and Parker, B. C. (1992). Cryopreservation of eukaryotic algae. *Virg. J. Sci.* **43**, 403–410.

Beijerinck, M. W. (1890). Kulturversuche mit Zoochlorellen, Lichenogoniden und anderen niederen Algen. *Bot. Ztg.* **48**, 725.

Beijerinck, M. W. (1891). Verfahren zum Nachwies der Säureabsonderung bei Mikrobien. *Zentralbl. Bakteriol., Parasitenkd. Infektionskr.* **9**, 781.

Beijerinck, M. W. (1893). Bericht über meine Kulturen niederer Algen of Nährgelatine. *Zentralbl. Bakteriol., Parasitenkd. Infektionskr.* **13**, 368.

Benecke, W. (1898). Über Kulturbedingungen einiger Algen. *Bot. Ztg.* **56**, 83.

Bold, H. C. (1942). The cultivation of algae. *Bot. Rev.* **8**, 69–138.

Bold, H. C. (1973). "Morphology of Plants," 3rd ed. Harper & Row, New York.

Bold, H. C., and Wynne M. J. (1985). "Introduction to Algae. Structure and Reproduction," 2nd ed. Prentice-Hall, Englewood Cliffs, NJ.

Cañavate, J. P., and Lubian, L. M. (1994). Tolerance of six marine microalgae to the cryoprotectants dimethyl sulfoxide and methanol. *J. Phycol.* **30**, 559–565.

Cooper, S., Battat, A., Marsot, P., and Sylvestre, M. (1983). Production of antibacterial activities by two Bacillariophyceae grown in dialysis culture. *Can. J. Microbiol.* **29**, 338–341.

Corbett, L. L., and Parker, D. L. (1976). Viability of lyophilized cyanobacteria (blue-green algae). *Appl. Environ. Microbiol.* **32**, 777–780.

Daily, W. A., and McGuire, J. M. (1954). Preservation of some algal cultures by lyophilization. *Butler Univ. Bot. Stud.* **11**, 139–143.

Douglas, S. E., Murphy, C. A., Spencer, D. F., and Gray, M. W. (1991). Cryptomonad algae are evolutionary chimaeras of two phylogenetically distinct unicellular eukaryotes. *Nature (London)* **350**, 148–151.

Drew, K. M. (1949). *Conchocelis*-phase in the life history of *Porphyra umbilicalis* (L.) Kütz. *Nature (London)* **164**, 748.

Druehl, L. D. (1988). Cultivated edible kelp. *In* "Algae and Human Affairs" (C. A. Lembi and J. R. Waaland, eds.), pp. 119–134. Cambridge Univ. Press, New York.

Falconer, I. R., Beresford, A. M., and Runnegar, M. T. C. (1983). Evidence of liver damage by toxin from a bloom of the blue-green algae, *Microcystis aeruginosa. Med. J. Aust.* **1**, 511–514.

Freifelder, D. (1964). Impurities in "pure" biochemicals. *Science* **144**, 1087–1088.

George, R. W. (1988). Products from fossil algae. *In* "Algae and Human Affairs" (C. A. Lembi and J. R. Waaland, eds.), pp. 305–333. Cambridge Univ. Press, New York.

Gherna, R., Pienta, P., and Cote, R., eds. (1989). "American Type Culture Collection Catalog of Bacteria and Phages," 17th ed. ATCC, Rockville, MD.

Gorham, P. R., and Carmichael, W. W. (1988). Hazards of freshwater blue-green algae (cyanobacteria). In "Algae and Human Affairs" (C. A. Lembi and J. R. Waaland, eds.), pp. 403–431. Cambridge Univ. Press, New York.

Guillard, R. R. L. (1975). Culture of phytoplankton for feeding marine invertebrates. In "Culture of Marine Invertebrate Animals" (W. L. Smith and M. H. Chanley, eds.), pp. 29–60. Plenum, New York.

Hallagraeff, G. M. (1993). A review of harmful algal blooms and their apparent global increase. Phycologia **32**, 79–99.

Harvey, W. H. (1836). Algae. In "Flora Hibernica" (J. T. Mackay, ed.), Part 3, pp. 157–254. William Curry & Co., Dublin.

Harvey, W. H. (1846–1851). "Phycologia Britannica," 4 vols. London.

Holm-Hansen, O. (1964). Viability of lyophilized algae. Can. J. Bot. **42**, 127–137.

Holm-Hansen, O. (1973). Preservation by freezing and freeze-drying. In "Handbook of Phycological Methods, Culture Methods and Growth Measurements" (J. R. Stein, ed.), pp. 195–205. Cambridge Univ. Press, New York.

Hwang, S.-W., and Hudock, G. A. (1971). Stability of Chlamydomonas reinharditii in liquid nitrogen storage. J. Phycol. **7**, 300–303.

Jackson, A. R. B., McInnes, A., Falconer, I. R., and Runnegar, M. T. C. (1984). Clinical and pathological changes in sheep experimentally poisoned by the blue-green alga Microcystis aeruginosa. Vet. Pathol. **21**, 102–113.

Jassby, A. (1988a). Spirulina: A model for microalgae as human food. In "Algae and Human Affairs" (C. A. Lembi and J. R. Waaland, eds.), pp. 149–179. Cambridge Univ. Press, New York.

Jassby, A. (1988b). Some public health aspects of microbial products. In "Algae and Human Affairs" (C. A. Lembi and J. R. Waaland, eds.), pp. 181–202. Cambridge Univ. Press, New York.

Jones, J. W., McFadden, H. W., Chandler, F. W., Kaplan, W., and Conner, K. H. (1983). Green algal infection in a human. J. Clin. Pathol. **80**, 102–107.

Kufferath, H. (1929). La culture des algues. Rev. Algol. **4**, 127–346.

Kuwano, K., Aruga, Y., and Saga, N. (1994). Cryopreservation of the conchocelis of Porphyra (Rhodophyta) by applying a simple prefreezing system. J. Phycol. **30**, 566–570.

Lee, J. J., and Soldo, A. T., eds. (1992). "Protocols in Protozoology." Society of Protozoologists, Lawrence, KS.

Lee, R. E. (1989). "Phycology," 2nd ed. Cambridge Univ. Press, New York.

Lembi, C. A., and Waaland, J. R., eds. (1988). "Algae and Human Affairs." Cambridge Univ. Press, New York.

Lewin, R. A. (1959). The isolation of algae. Rev. Algol. **3**, 181–197.

Lewin, R. A. (1977). Prochloron, type genus of the Prochlorophyta. Phycologia **16**, 217.

Lewis, J. G., Stanley, N. F., and Guist, G. G. (1988). Commercial production and applications of algal hydrocolloids. In "Algae and Human Affairs" (C. A. Lembi and J. R. Waaland, eds.), pp. 205–236. Cambridge Univ. Press, New York.

McGrath, M. S., Daggett, P.-M., and Dilworth, S. (1978). Freeze-drying of algae: Chlorophyta and Chrysophyta. J. Phycol. **14**, 521–525.

McHugh, D. (1987). Production, properties and uses of alginates. In "Production and Utilization of Products from Commercial Seaweeds," pp. 50–96. FAO/UN, Rome.

McLachlan, J. (1973). Growth media—marine. In "Handbok of Phycological Methods. Culture Methods and Growth Measurements" (J. R. Stein, ed.), pp. 25–51. Cambridge Univ. Press, New York.

McLellan, M. R. (1989). Cryopreservation of diatoms. Diatom Res. **4**, 301–318.

Miquel, P. (1893a). De la culture artificielle des diatomés. Introduction. *Le Diatomiste* **1**, 73–75.

Miquel, P. (1893b). De la culture artificielle des diatomés. Cultures ordinaires des diatomées. *Le Diatomiste* **1**, 93–99.

Miquel, P. (1893c). De la culture artificielle des diatomés. Culture articielle des diatomés marines. *Le Diatomiste* **1**, 121–128.

Miquel, P. (1893c). De la culture artificielle des diatomés. Cultures pures des diatomées. *Le Diatomiste* **1**, 149–156.

Miquel, P. (1893e). De la culture artificielle des diatomés. Culture des diatomées sous le microscope. *Le Diatomiste* **1**, 165–172.

Miura, A. (1975). *Porphyra* cultivation in Japan. In "Advances in Phycology in Japan" (J. Tokida and H. Hirose, eds.), pp. 23–304. Junk, The Hague.

Miura, A. (1979). Studies on the genetic improvement of cultivated *Porphyra* (laver). *Proc. Jpn. Soc. J. Symp. Aquacult.*, 7th, Tokyo, 1978, pp. 161–168.

Molisch, H. (1895). Die Ernährung der Algen. I. Süsswasseralgen. *Sitzungsber. Akad. Wiss. Wien, Math.-Naturwiss. Kl., Abt. 1* **104**, 783.

Molisch, H. (1896). Ernährung der Algen. II. Süsswasseralgen. *Sitzungsber. Akad. Wiss. Wien, Math.-Naturwiss. Kl. Abt. 1* **105**, 633.

Morel, F. M. M., Rueter, J. G., Anderson, D. M., and Guillard, R. R. L. (1979). Aquil: A chemically defined phytoplankton culture medium for trace metal studies. *J. Phycol.* **15**, 135–141.

Morris, G. J. (1976a). The cryopreservation of *Chlorella* 1. Interactions of rate of cooling, protective additive and warming rate. *Arch. Microbiol.* **107**, 57–62.

Morris, G. J. (1976b). The cryopreservation of *Chlorella* 2. Effect of growth temperature on freezing tolerance. *Arch. Microbiol.* **107**, 309–312.

Morris, G. J. (1978). Cyropreservation of 250 strains of Chlorococcales by the method of two-step cooling. *Br. Phycol. J.* **13**, 15–24.

Morris, G. J. (1981). "Cryopreservation. An Introduction to Cryopreservation in Culture Collections." Institute of Terrestrial Ecology, Cambridge, UK.

Mumford, T. F., Jr., and Miura, A. (1988). *Porphyra* as food: Cultivation and economics. In "Algae and Human Affairs" (C. A. Lembi and J. R. Waaland, eds.), pp. 87–117. Cambridge Univ. Press, New York.

Nerad, T. A., and Daggett, P.-M. (1992). Preservation of protozoa. Cryopreservation, drying and freeze-drying. General methods and definition of terms. In "Protocols in Protozoology" (J. J. Lee and A. T. Soldo, eds.), pp. A-56 to A-56.15. Society of Protozoology, Lawrence, KS.

Ohno, M., and Critchley, A. T. (Editors) (1993). "Seaweed Cultivation and Marine Ranching." Japan International Cooperation Agency, Yokosuka, Japan. 151 pp.

Okazaki, A. (1971). "Seaweeds and Their Uses in Japan." Tokai Univ. Press, Tokyo.

Palenik, B., and Haselkorn, R. (1992). Multiple evolutionary origins of prochlorophytes, the chlorophyll b-containing prokaryotes. *Nature (London)* **355**, 265–267.

Patwary, M. U., and van der Meer, J. P. (1992). Genetics and breeding of cultivated seaweeds. *Korean J. Phycol.* **7**, 281–318.

Perasso, R., Baroin, A., Qu, L. H., Bachellerie, J. P., and Adoutte, A. (1989). Origin of the algae. *Nature (London)* **339**, 142–144.

Prát, S., Dvořáková, J., and Baslerová, (M. 1972). "Cultures of Algae in Various Media." Publ. House Czech. Acad. Sci., Prague.

Price, N. M., Harrison, G. I., Hering, J. G., Hudson, R. J., Nivel, P. M. V., Palenik, B., and Morel, F. M. M. (1989). Preparation and chemistry of the artificial algal culture medium Aquil. *Biol. Oceanogr.* **6**, 443–461.

Pringsheim, E. G. (1946). "Pure Cultures of Algae." Cambridge Univ. Press, Cambridge, UK.

Provasoli, L., and Pintner, I. J. (1960). Artificial media for fresh-water algae: Problems and suggestions. *In* "The Ecology of Algae" (C. A. Tryon, Jr. and R. T. Hartman, eds.), Spec. Publ. 2, pp. 84–96. Pymatuning Laboratory, University of Pittsburgh.

Rabenhorst, L. (1864). "Flora Europaea. Algae aquae dulcis et submarinae," Sect. 1. Leipzig.

Rabenhorst, L. (1865). "Flora Europea. Algae aquae dulcis et submarinae," Sect. 2. Leipzig.

Rabenhorst, L. (1868). "Flora Europaea. Algae aquae dulcis et submarinae," Sect. 3. Leipzig.

Richmond, A. (1986). "CRC Handbook of Microalgal Mass Culture." CRC Press, Boca Raton, FL.

Richter, O. (1903). Reinkultur von Diatomeen. *Ber. Dtsch. Bot. Ges.* **21**, 493.

Sakagami, Y., Watanabe, T., Hisamitsu, A., Kamibayshi, Honma, K., and Manabe, H. (1982). Anti-ulcer substances from marine algae. *In* "Marine Algae in Pharmaceutical Science" (H. A. Hoppe and T. Levring, eds.). de Gruyter, Berlin. 99–108.

Schwimmer, M., and Schwimmer, D. (1955). "The Role of Algae and Plankton in Medicine." Grune & Straton, New York.

Shumway, S. E. (1990). A review of the effects of algal blooms on shellfish and aquaculture. *J. World Aquacult. Soc.* **21**, 65–104.

Smith, G. M. (1950). "Fresh-Water Algae of the United States," 2nd ed. McGraw-Hill, New York.

Stanier, R. Y., Kunisawa, R., Mandel, M., and Cohen-Gazire, G. (1971). Purification and properties of unicellular blue-green algae (order Chroococcales). *Bacteriol. Rev.* **35**, 171–205.

Stanley, N. (1987). Production, properties and uses of carrageenan. *In* "Production and Utilization of Products from Commercial Seaweeds," pp. 97–147. FAO/UN, Rome.

Starr, R. C., and Zeikus, J. A. (1993). UTEX—The culture collection of algae at the University of Texas at Austin. *J. Phycol.* **29**, Suppl., 1–106.

Stein, J. R., ed. (1973). "Handbook of Phycological Methods. Culture Methods and Growth Measurements." Cambridge Univ. Press, New York.

Stein, J. R., and Borden, C. A. (1984). Causative and beneficial algae in human disease conditions: A review. *Phycologica* **23**, 485–501.

Taggart, R. E., and Parker, L. R. (1976). A new fossil alga from the Silurian of Michigan. *Am. J. Bot.* **63**, 1390–1392.

Tappan, H. (1980). "The Paleobiology of Plant Protists." Freeman, San Francisco.

Tseng, C. K. (1981). Commercial cultivation. *Bot. Monogr. (Oxford)* **17**, 680–725.

Tsuru, S. (1973). Preservation of marine and fresh water algae by means of freezing and freeze-drying. *Cryobiology* **10**, 445–452.

van der Meer, J. P. (1988). The genetic improvement of algae: Progress and prospects. *In* "Algae and Human Affairs" (C. A. Lembi and J. R. Waaland, eds.), pp. 511–528. Cambridge Univ. Press, New York.

Yamamoto, I., and Maruyama, H. (1984). Inhibitory effects of dietary seaweeds (*Undaria, Porphyra, Laminaria*) on the growth of spontaneous mammary carcinoma in C3H mice. *Proc. Jpn. Cancer Assoc., 43rd Annu. Meet.,* Fukuoka, Abstract.

Yamamoto, I., and Maruyama, H. (1985). Effect of dietary seaweed preparations on 1,2-dimethyhydrazine-induced intestinal carcinogenesis in rats. *Cancer Lett.* **26**, 241–251.

CHAPTER 4

Preservation and Maintenance of Eubacteria

L. K. Nakamura

Eubacteria

Diversity

As indicated in Table 1, bacteria are morphologically, physiologically, biochemically, and genetically diverse groups of organisms. The morphologically simple forms of these organisms separate into three general categories, namely, cocci, rods, and spirals. However, subgroupings exist within each category. For example, cocci can be arranged in chains, in cubical packets, or in random agglomerations. The rods can be large or small, have peritrichous or polar flagella or none at all, produce endospores, or extend into long chains. The spirals can corkscrew a few turns or many.

Not all bacteria are morphologically simple. A large group of bacteria display irregular outlines; these are called "coryneforms" and include genera such as *Corynebacterium, Arthrobacter, Microbacterium, Brevibacterium, Cellulomonas,* and others (Buchanan and Gibbons, 1974). Other complex forms include sheathed bacteria (*Leptothrix, Streptothrix, Lieskeella, Phragmidiothrix, Crenothrix,* and *Clonothrix*); appendaged (prosthecae, holdfasts) bacteria (*Hyphomonas, Caulobacter, Gallionella,* and *Planctomyces*); and bacteria (Myxobacterales) that aggregate to form often brightly colored and usually macroscopically sized fruiting bodies constructed of slime and cells (Buchanan and Gibbons, 1974).

TABLE 1
Diversity Exhibited by Bacteria

Characteristics	Diversity
Growth temperature	Psychrophiles grow at 5–10°C; mesophiles grow at 25–45°C; thermophiles grow at 50–80°C; thermal vent organisms can grow at 100°C or higher
Oxygen requirement	Obligate aerobes; facultative anaerobes; obligate anaerobes
Microscopic morphology	Cocci occur singly, in chains or in packets; rods occur in chains or singly; spirals occur singly or in short chains; size varies from <1.0 to 10 μm or so; endospores may be produced; some are motile; coryneforms; appendages formed by some
Macroscopic morphology	Colonies vary in size, texture, translucency, perimenter characteristics; fruiting bodies formed
Metabolism	Heterotrophic (fermentative, oxidative); autotrophic (chemoautotrophic, photoautotrophic); nitrogen fixing
Staining reaction	Gram positive; gram negative; gram variable; acid fast
G + C contents (mol%)	Range from 28–30 for *Spirillum linum* to 70–80 for *Micrococcus luteus, Mycobacterium smegmatis,* and *Streptomyces* spp.

Physiologically, bacteria exhibit the widest range of possibilities. They exist as mesophiles, thermophiles, or psychrophiles; some fastidiously require narrow temperature ranges for growth, whereas others tolerate wide ranges. The oxygen demand ranges from strictly aerobic, to facultatively anaerobic, and to strictly anaerobic, where even a few molecules of oxygen can be toxic to the organism. Metabolically, these organisms can be heterotrophic or autotrophic. Bacteria can be saprophytic, pathogenic (*Pasteurella, Salmonella*), parasitic (*Bdellvibrio*), or symbiotic (*Rhizobium, Bradyrhizobium*). Some of these characteristics play important roles in the preservation and maintenance of the organisms.

Classification

Classification is the orderly positioning of organisms into taxonomic groups on the basis of similarities or relationships. One of its purposes is to organize a diverse array of individuals, such as the bacteria, into an orderly arrangement. A hierarchical ranking system is used in bacterial classification to take into account the varying levels of similarities or relationships. For example, all prokaryotic organisms are placed in the Kingdom Prokaryotae. Divisions, classes, orders, families, genera, and species are smaller, nonoverlapping subsets of the kingdom. A specific example of taxonomic ranking is shown in Table 2. Bacterial classification is presently designed mainly for identification rather than for showing evolutionary relationships.

SPECIES

Evolution implies a relatedness among all organisms. Higher animals and plants can be clearly divided into species, strains of which interbreed successfully and produce fertile offspring. In contrast, bacteria reproduce asexually; thus, they have no natural species boundaries. Bacterial species may be regarded as collections of strains that share many common characteristics and are distinctly different from strains of other species. In bacteriology, a type strain is designated. This strain serves as the permanent example of the species; it has great importance for classification at the species level because "a species consists of the type strain and all other strains that are considered to be sufficiently similar to it as to warrant inclusion within the species" (Staley and Krieg, 1984). Because the type strain is such an important factor in bacterial taxonomy, its characteristics must be kept as unvarying as possible by use of proper preservation meth-

TABLE 2

Taxonomic Ranking of *Bacillus subtilis*

Rank	Example
Kingdom	Prokaryotae
Division	Firmicutes
Class	Firmibacteria
Order	Eubacteriales
Family	Bacillaceae
Genus	*Bacillus*
Species	*subtilis*

ods. Because this concept of species involves subjective judgment, some species, not surprisingly, are phenotypically and genetically more diverse than others.

However, increasing use of emerging methods is allowing taxonomists to propose new species with more objectivity than ever before. One of these methods involves the analysis of nucleic acids. A feature of bacterial DNA having taxonomic values is the guanine-plus-cytosine (G + C) content. Among bacteria, the G + C content ranges from about 25 to 75 mol% (Johnson, 1984). Closely related species have strains with very similar G + C values. However, organisms with similar G + C contents may not necessarily be closely related, because G + C determination does not take into account the linear arrangement of the nucleotides in the DNA molecule.

DNA–DNA hybridization is widely used to define species boundaries; presently, the consensus is that 70% or higher complementarity between DNAs of two separate individuals indicates conspecificity (Johnson, 1984). Other procedures that have been used to differentiate taxa include numerical taxonomy (Sneath and Sokal, 1973); chemotaxonomy, which includes analysis of cell wall composition (Kandler and Schleifer, 1980; Keddie and Bousfield, 1980), of lipid composition (Shaw, 1975; Lechevalier, 1977; Kates, 1978), of isoprenoid quinones (Jeffries, 1969; Yamada et al., 1976; Collins and Jones, 1981), of cytochrome composition (Jones, 1980), of amino acid sequences of various proteins (Doolittle, 1981), of protein profiles (Kersters and De Ley, 1975), of fatty acid profiles (Sasser, 1990), and of enzyme composition (Selander et al., 1986); serology; and genetics, namely transformation (Bøvre, 1980). Sequencing of ribosomal RNA or DNA has done much to elucidate phylogenetic relationships of various bacterial taxa. However, the information has not been incorporated into the formal classification scheme for bacteria.

GENUS

All species are assigned to a genus. Thus, bacterial nomenclature conforms to the binomial system of Linneus, whereby organisms are designated by a combined genus and species name. Although bacterial genera are well-defined groups and clearly distinguishable from each other, there is no general agreement on the definition of a bacterial genus, and considerable subjectivity is encountered at the genus level. The application of DNA and ribosomal RNA sequencing allows for increased objectivity at the genus level.

HIGHER TAXA

The relationships at higher classifications levels are less certain than those at the genus and species levels. Frequently, there is little basis for assignment of taxa at the family and higher levels. Therefore, one of the present classification schemes subdivides bacteria into four divisions (Division 1, Gracilicutes; Division II, Firmicutes; Division III, Tenericutes; and Division IV, Mendosicutes) that are further subdivided into classes (Murray, 1984). There is no general agreement about this or any other arrangement of divisions and classes. Further subdivision into order and family becomes less certain. However, as more is learned about the genetic relationships of bacteria by RNA and DNA sequencing techniques, meaningful familial and ordinal placements will ensue.

Characterization

Presently, for diagnostic purposes, bacteria are described by characteristics that are relatively easy to recognize (Table 3). These include macroscopically or microscopically visible features such as shape, size, motility, staining reactions, production of spores, flagellation pattern, capsule formation, and colonial morphology; production of characteristic metabolic substances; fermentation of carbohydrates to produce acid and/or gas; nutrition, including growth requirements and ability to utilize various substrates for energy and nitrogen source; ecological niche (saprophytic, symbiotic, parasitic, pathogenic); and chemical composition (peptidoglycan, total fatty acid, protein profile, nucleic acid).

Industrial Importance

Industrially, the importance of bacteria can be negative or positive. On the negative side, bacteria as contaminants can cause havoc in the fermentation industry. Product yield is decreased by bacterial contaminants through competition for nutrients or by actual degradation. Therefore, considerable efforts must be taken to ensure the sterility of all equipment and chemicals that are used in the process. As spoilage agents, these organisms cause millions of dollars worth of damage in the food industry. In food processing, where heat sterilization is used to extend shelf-life, every attention must be given to the invactivation of the spores of both facultatively and obligately anaerobic bacteria. An especially undesirable spore-forming, food spoilage organism is the anaerobe, *Clostridium botulinum*, because it concomitantly destroys the quality of the food product and produces the potent botulism-causing toxin. Aside from food poisoning, bacteria cause many important diseases that are expensive to treat. The need to find drugs and means to treat bacterial diseases has spurred the creation of a vast pharmaceutical industry.

On the positive side, bacteria are used in many industrial processes. Fermentation industries use bacteria to produce organic acids (e.g., lactic

TABLE 3
Methods for Characterizing Bacteria

Method	Traits examined	Examples of traits noted
Morphology		
Macroscopic	Colony	Size; color; texture; translucency; edge characteristics
Microscopic	Cell shape, size	Coccus; rods; spirals
	Structures	Spores; flagella; capsule; holdfast; prothecae
	Staining reactions	Gram; acid-fastness
Metabolic products	From carbohydrates and related compounds	Acids; gas (carbon dioxide and hydrogen); alcohols; aldehydes; ketones
	From nitrogenous compounds	Indole; nitrogen gas; ammonia
Nutritional requirements	Trace materials	Minerals; vitamins; other factors
	Nitrogen sources	Proteins; amino acids; inorganic compounds; atmospheric nitrogen
	Carbon sources	Organic compounds; carbon dioxide
	Energy sources	Organic and inorganic compounds; light
Physiology	Oxygen requirement	Anaerobic; facultatively anaerobic; strictly aerobic
	Growth temperature	Psychrophiles (0–15°C); mesophiles (25–45°C); thermophiles (55°C+)
	Metabolic type	Oxidative or fermentative
	Autotrophs	Chemoautotrophic; photosynthetic
	Enzyme activities	Catalase; oxidase; carbohydrase; lipase; protease; nitrate reductase

(Continues)

TABLE 3 (*Continued*)

Method	Traits examined	Examples of traits noted
Ecology	Special environment	—
	Relation to other organisms	Saprophytic; parasitic; symbiotic
Chemotaxonomic	Whole cell	Protein profile; enzyme profile; fatty acid profile; lipid composition; isoprenoid; quinones; cytochrome composition; amino acid sequence of proteins
	Specific structures	Cell wall; genetic materials (G + C content of DNA; DNA and RNA sequences; plasmids)
Serology	Cell surface antigens	Cell wall lipopolysaccharides; surface layer proteins
	Other components	Spores; flagella; capsular materials
Genetic methods	DNA hybridization, transformation	Nucleic acid sequencing

acid by *Lactobacillus delbrueckii* and acetic acid by *Acetobacter aceti*), alcohols (butanol by the anaerobe *Clostridium acetobutylicum*), enzymes (amylases by *Bacillus subtilis*), antibiotics (polymyxin by *Bacillus polymyxa*), polysaccharides (xanthan by *Xanthomonas campestris* and dextran by *Leuconostoc mesenteroides*), vitamins (vitamin B_{12} by *Lactobacillus leichmanii*), and a host of other substances. Recent work suggests that marine bacteria may be sources of novel antibiotics and other bioactive agents. Foods such as cheeses (lactobacilli, propionibacteria), yoghurt (lactobacilli), vinegar (acetobacteria), pickles (lactobacilli), sauerkraut (lactobacilli), olives (lactobacilli), sausage (lactobacilli), and soy sauce (lactobacilli) derive their unique flavors entirely or partly from the activity of the indicated bacteria. Genetically engineered or recombinant organisms such as *Escherichia coli* and *B. subtilis* are used to produce chemicals not normally produced by the microorganisms, examples being mammalian hormones and enzymes. Bacteria such as *E. coli* are also important agents for the handling of genetic materials for carrying out genetic manipulations. Bacteria are used as assay organisms; for example, *Micrococcus luteus, E. coli, Pseudomonas aeruginosa,* and *Lactobacillus* spp. are used for assaying vitamins, antibiotic activity, and amino acid.

In agriculture, benefits accrue from the ability of bacteria to fix nitrogen symbiotically (*Rhizobium* and *Bradyrhizobium*) or nonsymbiotically (*Azotobacter, Derxia, Bacillus,* and others). Consortia of bacteria are involved in the decomposition of agricultural waste into harmless chemicals or into useful compost. Lactic acid bacteria are the mainstay in the ensilaging process, which is vital for preservation and enhancement of the quality of farm animal fodder. Bacteria can also have impact on agriculture by their unwanted effects, such as denitrification (Alexander, 1961) and causation of diseases in crops and farm animals. The genera *Erwinia, Xanthomonas,* and *Pseudomonas* are implicated in diseases that cause significant destruction of crops. Bacteria such *Brucella abortus, Bacillus anthracis, Staphylococcus aureus, Pseudomonas mallei,* and *Hemophilus suis* cause abortion, anthrax, mastitis in cows, glanders in horse, and swine influenza, respectively (Merchant, 1950).

Without bacteria, large to medium-sized municipalities would find it difficult to purify their water supplies and to process waste. Successful decomposition of organic matter in the water depends on the formation of "flocs," which mainly contain the bacteria *Zoogloea ramigera* (Crabtree *et al.,* 1965).

Methodologies

Culture Conditions

GROWTH CONDITIONS

Because bacteria are diverse organisms, generalizations cannot be made regarding pH values, media, and incubation temperatures that will support optimum growth. Each organism to be grown must be considered individually and conditions must be selected to suit it. Many organisms, however, will grow at pH values near neutrality, at about 30°C, in a medium containing an energy source such as glucose and an organic nitrogen source such as peptone. Then there are extremes. The autotrophs may not tolerate any organic compounds and require low pH (*Nitrobacter*), strict thermophiles require temperatures higher than 55°C (*Bacillus acidocaldarius*), obligate anaerobes (*Clostridium, Butyrivibrio, Bacteriodes*) do not tolerate the presence of any oxygen, parasitic organisms (*Bdellvibrio*) need suitable hosts to grow on, halophiles and marine organisms grow only in media containing high salt concentrations, recombinant organisms need to grow in media formulated to maintain inserted characteristics and nutritionally fastidious organisms will require a number of biochemicals such as vitamins, blood components, or specific organic compounds (*Lactobacillus* spp., several pathogens).

PHYSIOLOGICAL STATE

In general, cultures to be preserved should be grown under optimum conditions into late log or early stationary phase. Optimum growth conditions should elicit the best titer to ensure survival during preservation. Where applicable, organisms should be grown on media that will elicit sporulation because spores survive preservation conditions well. Once growth has occurred, there are several different methods for preserving bacteria.

Preservation

CONTINUOUS SUBCULTURE

One of the oldest and most traditional methods for preserving bacterial cultures is continuous subculturing. Organisms have to be grown on optimum media; some species require transfer after days or weeks, whereas others may be transferred after several months or years.

Subculturing has many disadvantages, the most notable of which are listed below:

1. Change of characteristics. Subculturing can lead to change of characteristics, i.e., characteristic may be lost, reduced, or intensified. Changes probably occur most frequently among strains where the intervals between transfers are short.
2. Mislabeling. Cultures may be labeled with the wrong name or number. Labels may become distorted and unrecognizable.
3. Contamination. This occurs frequently, especially when large numbers of cultures are involved and the concentration of the person doing the transfers lags. The use of good technique and transfer hoods, and frequent rest by the personnel doing the transfers, would help eliminate the problem.
4. Inoculation with the wrong organism. This error happens when large numbers of cultures have to be transferred. The work is tedious and not conducive to maintaining concentration. Organisms may be inoculated into the wrong tube or several organisms may be put into the same tube.

5. Loss of cultures. This situation occurs from time to time and is probably more common with the more "delicate" organisms. Temperature fluctuations in incubation or refrigeration equipment contribute to the possibility of loss. Extreme fluctuation in temperature can lead to loss of large numbers of cultures. Cultures also may be lost because they are not transferred.
6. Storage space. Large numbers of test tube cultures demand very large storage and incubation space.

Because of the many disadvantages associated with the method, continuous subculturing is obviously not the best method for preserving or maintaining bacterial cultures.

LYOPHILIZATION

This procedure obviates many of the disadvantages associated with the subculturing procedure. One of the most significant advantages of this procedure is reduction of variability because of the infrequency of producing new cultures, and the length of survival time. Once properly grown, most bacteria, including recombinants and the "extremophiles" (such as thermophiles, psychrophiles, obligate anaerobes, acidophiles, alkalophiles, halophiles, marine organisms, and others), can be preserved by lyophilization. Furthermore, because cultures are preserved in small ampules, large numbers can be stored in relatively small spaces; these ampules are light and easy to ship.

Principles of Lyophilization. Lyophilization is a freeze-drying process; water is removed by sublimation from frozen cellular material (Rowe and Snowman, 1976; Rey, 1977). Rapid freezing and use of suitable suspending medium protect the cellular structures from damage. Reconstitution of lyophilized materials is rapid because the original shape is preserved and a very large surface area is available. Three steps are involved in the freeze-drying process: prefreezing to provide a solidly frozen starting material; primary drying, which removes most of the water; and secondary drying, which removes bound water.

Both the freezing method and the final temperature of frozen material can affect how well the samples can be freeze-dried. Generally, freezing damages living matter by the formation of ice crystals, and by the increase of electrolyte concentrations resulting from removal water by the freezing process. Presumably, removal of water from proteins and nucleic acids causes the damage. Rapid freezing of cell suspensions elicits formation of small ice crystals, whereas slow freezing rates promote formation of large ice crystals. Frozen samples with small ice crystals are harder to freeze-dry, but they suffer minimal damage due to ice crystal formation. To achieve effective lyophilization, suspensions must be cooled until all eutectic mixtures (high concentrations of solutes resulting from removal of water) are frozen.

The eutectic temperature of a suspension can be determined experimentally by measuring the electrical resistance as the suspension is being frozen. The resistance will remain low until the eutectic temperature in attained, whereupon the resistance will increase sharply (Simione and Brown, 1991).

Primary drying involves removal of moisture from the frozen sample by sublimation, which is facilitated by imposition of a vacuum. Therefore, a large-capacity vacuum pump is an essential element of the freeze-drying system. A moisture trap or condenser is needed to prevent water vapor from entering the pump. The condenser temperature must be lower than that of the frozen cell suspension for sublimation to occur. The rate of sublimation is related to the vapor pressure differential between the condenser and frozen sample. To increase the vapor pressure differential, heat can be applied to the frozen sample at a level that does not warm the frozen sample beyond the capacity of the evaporative cooling. In some systems, ambient heat is used, and usually evaporative cooling can maintain the frozen state of the sample. More sophisticated systems allow control of heat input, which maximizes drying without causing damage to the sample. Monitoring condensable and total pressure (using pressure gauges designed to measure both condensable and noncondensable gases) is an effective way of determining the end point of primary drying.

Bound water that is not removable by sublimation must be desorbed by application of heat. Secondary drying is carried out under vacuum and requires a moisture trap and warming of the sample. A residual moisture content of 1% or less is desired to ensure good shelf life, although studies have shown that 2–3% residual moisture is required to maintain the stability of lyophilized microorganisms.

Practice of Lyophilization by the Manifold Method. Cells grown in broth or on slants are harvested and suspended in a medium that will protect the cells against the ravages of freezing and drying. Because survival is directly related to the initial cell titer, a relatively dense suspension should be used to ensure optimal results. With nonsporulating organisms, a titer of about 10^{10} cells/ml is desirable. In general, Gram-positive bacteria survive freeze-drying better than Gram-negative organisms (MacKenzie, 1977). When spore-forming organisms such as *Bacillus Clostridium, Sporolactobacillus,* or *Sporosarcina* are lyophilized, it is preferable to lyophilize the spores rather than the vegetative cells. Anaerobic organisms must be handled using procedures that will prevent their exposure to oxygen.

The survivability of lyophilized organisms is greatly affected by the suspending medium. In addition to ensuring survivability, the suspending medium should allow easy recovery of viable organisms. Lapage *et al.* (1970) observed that the addition of 7.5% glucose greatly improved the survivability of organisms suspended in nutrient broth. Further improvements were noted when most of the nutrient broth was replaced by sterile bovine or equine serum. The glucose supplementation prevents removal of more than 99% of the moisture. Studies showed that total removal of water resulted in death of the organisms; approximately 1% of water must be retained to ensure the viability of the lyophilized preparations. Although possibly less effective than serum, skim milk can also be used a suspending medium. In situations when cost is a consideration, skim milk would be the medium of choice.

Lyophilization can be carried out on a simple apparatus (Haynes *et al.,* 1955) consisting of a manifold, to which glass ampules can be attached by means of vacuum-tubing nipples, a cold-finger trap, and a vacuum pump. In the first step, small amounts of the suspension are delivered to sterile glass ampules, which are plugged with sterile cotton. The ampules are then attached to the nipples of the manifold of the lyophilization apparatus and are immersed in an ethylene glycol (50%) dry-ice bath at $-40°C$ to freeze the cell suspension. Evacuation is started and the temperature of the bath is allowed to rise on its own accord. At this point, the vacuum measures from 100 to 200 millitorrs. When the temperature reaches about $-5°C$, drying proceeds rapidly. When inspection shows that the preparations appear completely dried (vacuum reading ranges from 30 to 60 millitorrs), the ampules are removed from the freezing bath and drying is allowed to continue at room temperature for about 30 to 45 minutes. The ampules are then sealed with an oxygen-gas torch under vacuum (less than 10 millitorrs). If desired, ampules can be tested for vacuum with a high-frequency tester.

Other methods are also available. For example, at the American Type Culture Collection (ATCC), three other methods are used:

1. Component freeze-dryer. Samples are lyophilized in cotton-plugged inner vials that are then sealed in outer vials under vacuum.
2. Commercial freeze-dryer. Samples are dried with a commercial freeze-dryer in vial-in-vial containers.
3. Preceptrol. Material is lyophilized in glass serum vials sealed with a rubber stopper and metal cap.

For specific details concerning protocols for the apparati used, see the work by Simione and Brown (1991).

Experience shows that viability is best maintained when lyophilized cultures are stored at 4°C. Lyophilized preparations should be stored in the dark because there are indications that light, especially fluorescent light, is harmful to freeze-dried organisms. Published reports indicate that lyophilized cultures can retain their viability at least 15 years (Lapage *et al.*, 1970). At the ARS Culture Collection, cultures that were lyophilized in the 1950s are still viable.

CRYOPRESERVATION

For some bacteria and organisms that are not amenable to preservation by lyophilization, an alternative is cryopreservation, i.e., preservation at -100 to $-273°C$. Cryogenic storage temperatures now commonly used for live cells are those of liquid nitrogen ($-196°C$) and liquid nitrogen vapor for higher storage temperature.

Principle. Cryopreservation is the freezing and storage of cells at very low temperatures (Mazur, 1970). Freezing and subfreezing temperatures cause to occur in living cells many changes that may be detrimental (Lasalle, 1974; Mazur, 1977). These include the formation of ice crystals; exosmosis; increased solubility of gases; dehydration; increased concentration of electrolytes, colloids, salts, carbohydrates, lipids, and proteins; lowering of pH; changes in heat and electric conductivity; decreased activity of some enzymes; increased activity of other enzymes; accumulation of intermediate metabolites; reduced intermolecular space; increased molecular contacts; disruption of weak hydrogen bonds; breakdown of emulsions; folding and distortion of large molecules; loss of cell membrane integrity; cellular invasion by toxic and mutagenic salts; eutectic concentration of all solutions; solidification; and immobilization of all molecules. As aqueous suspensions of living cells are cooled, liquid water around the cells begins to turn to ice, resulting in increasing the concentrations of solutes outside the cells. Because a differential in osmotic pressure is set up, water leaves the cells and continues to do so as long as the imbalance in salt concentration remains. This is a drying process, the rate and extent of which depend on the rate of cooling and the permeability of the cells.

Removal of too much water from the cells causes increase of internal solute concentrations, which may be harmful. However, too much water in the cell leads to intracellular damage due to ice crystal formation. The delicate balance of the two phenomena is maintained by carefully controlling the cooling rate when freezing cells, and warming the cells as rapidly as possible during the thawing process.

Because both solute concentration and ice formation contribute to cell damage, the cooling process should neither be too fast nor too slow. In practice, a rate of $10°C/minute$ is suitable for most living cells.

Cryoprotective agents can be added to cell suspensions to help minimize the damage occurring during freezing. Good cryoprotective agents have the following characteristics (Meryman, 1971; Fahy, 1986): (1) nontoxic to cells, (2) penetrate the cell membrane easily, and (3) bind either the electrolytes that increase in concentration during freezing or the water molecules to delay freezing. Two of the most common and effective cryoprotective agents are glycerol and dimethyl sulfoxide (DMSO).

Storage temperature can be critical for survival of frozen cells. Ice crystal formation continues below the freezing temperature of most frozen suspensions. Therefore, it is important for some cell systems that they be held below a critical temperature, which is −130°C (Meryman, 1966).

Most of the undesirable effects of freezing and thawing may be prevented or minimized for cryopreservation of live cells through a few simple basic steps.

1. Cells are grown into mid or to late log phase at the proper temperature on the best medium.
2. Cell count is adjusted to $(2-6) \times 10^6$ cells/ml. Glycerol or DMSO is added and mixed immediately to final concentrations of 5%. This level of cryoprotectant has been shown to be adequate for bacterial cells (Lapage *et al.*, 1970).
3. The suspension is dispensed into cryogenic ampules, sealed, and then held for 30 minutes at 30°C in a 0.05% methylene blue solution to allow osmotic equilibration between the cells and the cryoprotective agent, and to allow the dye to seep into improperly sealed ampules. If ampules are to be stored by immersion into the liquid phase of the liquid nitrogen, heavy-walled borosilicate ampules should be used and testing for leaks would be a must. Testing for leaks is essential because it obviates the potential hazard of improperly sealed ampules exploding when they are removed from the liquid nitrogen and rapidly brought to room temperature. The hazard can be avoided by storing the ampules in the vapor phase of the liquid nitrogen refrigerator. Commercial sealing equipment is available and preferably should be sealed by the pull-seal instead of by the tip-seal method because the occurrence of pinhole leaks is minimized.
4. The temperature is lowered at a rate of 1 to 3 degrees/minute to −30°C, followed by a more rapid rate of 15 to 30 degrees/minute to −100°C or lower; the sample is then transferred to −196°C liquid nitrogen. A number of programmable commercial instruments are available for controlled freezing of cell suspensions. These generally operate on a differential thermocouple principle. Although precisely controlled freezing may be essential for selected organisms and to ensure reproducibility of preservation from one time to another, many cells may be successfully frozen by simply placing the ampules in a dry ice chest or mechanical refrigerator set at −65°C for several

hours. Freezing under these conditions usually provides an adequate rate of freezing.

Generally, recovery is accomplished by removing the ampules as rapidly as possible from the liquid nitrogen refrigerator and plunging it immediately into a water bath at 37–40°C. Thawing should occur as rapidly as possible; moderate agitation usually effects melting of the ice in 40 to 60 seconds. Studies show that bacteria can retain up to 95% viability after 10 years of storage in liquid nitrogen.

Cryogenic temperatures of −135 to −145°C can also be attained in mechanical freezing units. These offer the advantage of not having to purchase liquid nitrogen, which can be a costly item, and eliminates the danger of leakage. Liquid nitrogen can be hazardous because it has a large expansion factor; leaked nitrogen has the capacity to displace all the air in a small room and can cause asphyxiation of workers in that room. The disadvantage of mechanical freezers is shutdowns that may occur with power outages. Unless discovered quickly, rises in temperature can be significant enough to result in loss of the preserved materials.

It should be noted that volumes 1 and 2 of "Bergey's Manual of Systematic Bacteriology" (Krieg and Holt, 1984; Sneath *et al.*, 1986) have for each of the genera included a "maintenance and preservation" section that lists the most appropriate method for preservation of the organism.

OTHER PRESERVATION METHODS

When lyophilization and liquid nitrogen equipment are not available, other methods employing the same basic principles of preservation are available. The methods can be separated into three general classes, namely, reduced metabolism, drying, and freezing.

Reduced Metabolism. Frequency of subculturing can be reduced by using a medium with minimal nutrients, by layering with mineral oil, and by storage at low temperatures. Organisms growing on nutritionally complete agar slants may also be covered with at least an inch of sterile mineral oil and stored at refrigeration temperature. Bacteria will survive many years under these conditions. This procedure is especially effective for preservation of fungi. Fully grown stab cultures sealed with a waxed cork and stored at 4°C have remained viable for many years (Lapage *et al.*, 1970).

Sporulating bacteria may be preserved by drying in sterile soil. Suspensions of the sporulating organisms are mixed with sterile soil, allowed to dry at room temperature, and then stored at refrigeration temperature. Sterile sand, silica gel, and many other inert materials have also been used successfully.

Drying. This method has been used for preservation of many kinds of organisms; some withstand the procedure and others do not. Cells or spores can be applied to a string (Annear, 1962), paper (Lapage et al., 1970), or mixed with gelatin (Stamp, 1947) and dried in a desiccator with or without vacuum. Cells preserved by these methods can survive several months of storage and are convenient to transport.

Freezing. This method is a variation of preservation in liquid nitrogen, and thus shares the same principles of operation, i.e., the rate of cooling should be low until $-20°C$ is reached and then as rapid as possible until the storage temperature is reached, electrolytes should be kept at a minimum, and cryoprotectants may be added to protect against damage.

Many bacteria will survive for extended periods of time at 4°C, especially if the test tubes are sealed with wax before refrigeration. Some bacteria, such as *Neisseria gonorrhoeae* and *Haemophilus* spp., cannot survive storage at this temperature.

Storage can be carried out successfully at $-20°C$, but care must be taken to avoid formation of eutectic mixtures. It is preferable to freeze at $-70°C$, but even at this temperature some loss of viability occurs.

When freeze-drying or liquid nitrogen freezing does not work or is not available, satisfactory results may be obtained with the L-drying method, a process that was first described by Annear (1962, 1964). In this procedure, small drops of culture are placed on cotton fibers and are then dried and sealed under vacuum. Temperature measurements indicate that no freezing occurs, hence drying takes place directly from the liquid state, hence the name L-drying. An essential step of the process is control of the degassing and moisture removal phase to avoid bubbling and, later, freezing of the culture. Published results indicate that this technique has been successful for bacteria, yeasts, and viruses. A modification of this process is used by the National Collection of Industrial Bacteria (Aberdeen, Scotland) and by the Institute for Fermentation (Osaka) for preservation of cultures that cannot be lyophilized or frozen in liquid nitrogen.

EVALUATION OF VIABILITY AND STABILITY

It is well known that changes in character occur commonly in cultures that are maintained by continued subculturing. The extent of changes associated with freeze-drying has not been systematically documented. However, some studies have shown that the freeze-drying process may cause mutations (Tanaka et al., 1979) and may select for organisms that are resistant to the various detrimental effects of freezing, hence enabling them to survive.

Whenever possible, a systematic protocol should be practiced to evaluate the stability of key characteristics of important organisms, such as type strains, strains used in production of desirable products, and strains used as assay organisms. Many culture collections make it a practice to send strains back to the original depositor to verify that the organism has retained its original characteristics.

If possible, a regular regimen should be arranged for checking the viability of preserved cultures. Preservation processes, be they lyophilization, drying, or freezing, stress organisms. Therefore, revival of organisms from the "resting" state may require complete media. It has also been found that Gram-positive bacteria survive lyophilization better than Gram-negative bacteria (MacKenzie, 1977).

Generally, the survival rate of lyophilized cultures is related to the initial numbers of cells; the higher the initial count in cell suspension, the higher the survival rate that can be expected. Furthermore, recovery from small numbers of organisms from the dried state can be very slow with certain organisms, especially autotrophs and anaerobes, and, consequently, the time required for obvious growth may increase. The time required for obvious growth for most organisms is 2–3 days; however, if small numbers of cells were processed, the time for obvious outgrowth may be as long as 10–14 days. Sometimes, 1–2 months may elapse before any outgrowth appears (Lapage *et al.*, 1970).

The ability to survive lyophilization is related to the age of the culture. Cultures that are too young or too old are more susceptible to low-temperature stress than are cultures that are in late log or early stationary growth phase. Growth conditions also influence resistance to the rigors of lyophilization. For example, aerated and shaken cultures are more resistant than are static or agar-grown cultures.

Although determination of viability is qualitative (the culture either grows or does not grow), viability may be quantitated by recording cell counts before lyophilization and at intervals during storage. Counts can be estimated by the standard dilution method. In a simpler method (Simione and Brown, 1991), the lyophilized sample is rehydrated in 0.3–0.5 ml of broth medium and the rehydrated suspension is added to 5 to 6 ml of the broth medium. Then, 0.1 ml of the suspension is inoculated at three spots on the edge of an agar plate. The plates are tilted to allow the drops to run across the agar surface, forming three streaks. The broth and plates are incubated and both are inspected for growth. Preparations with fewer than 100 colonies/streak are considered unsatisfactory.

Eubacteria Repositories

American Type Culture Collection (ATCC)
12301 Parklawn Drive
Rockville, MD 20582-1776

Telephone: 1-301-881-2600
Telefax: 1-301-231-5826

Collection des Bacteries de l'Institut Pasteur
25 rue du Docteur Roux/B.P. 52
75724 Paris Cedex 15
Telephone: 45 68 87 75
Telefax: 40 61 30 07

Deutsche Sammlung von Mikrooganismen und Zellkulturen
GmbH (DSM), Mascheroder Weg 1b
D-38124 Braunschweig, Germany
Telephone: 05 31/26 16-0
Telefax: 05 31/26 16-418

Japan Collection of Microorganisms
Riken, Wako-shi, Saitama 351-01, Japan
Telephone: 0484-62-1111 ext. 5100
Telefax: 0484-64-5651

Institute of Fermentation, Osaka (IFO)
17-85, Juso-honmachi 2 Chome, Yodogawa-ku
Osaka 532, Japan
Telephone: 06-302-7281
Telefax: 06-300-6814

National Collection of Type Cultures (NCTC)
Central Public Health Laboratory
Colindale Avenue
London, NW9, UK
Telephone: 01-205-7041
Telefax: (44) 181-2007874

Source for Sterile Bovine Serum

Colorado Serum Company
4950 York Street
Denver, Colorado 80216

References

Alexander, M. (1961). *In* "Introduction to Soil Microbiology," pp. 293–308. Wiley, New York.
Annear, D. I. (1962). Recoveries of bacteria after drying on cellulose fibers. A method for routine preservation of bacteria. *Austral. J. Exp. Biol. Med. Sci.* **40,** 1–8.
Annear, D. I. (1964). Recoveries of bacteria after drying in glutamate and other substances. *Austral. J. Exp. Biol. Med. Sci.* **42,** 717–722.
Bøvre, K. (1980). Progress in classification and identification of Neisseriaceae based on genetic affinity. *In* "Microbial Classification and Identification" (M. Goodfellow and R. G. Board, eds.), pp. 55–72. Academic Press, New York.

Buchanan, R. E., and Gibbons, N. E. (1974). *In* "Bergey's Manual of Determinative Bacteriology," 8th Ed. Williams & Wilkins, Baltimore, Maryland.

Collins, M. D., and Jones, D. (1981). The distribution of isoprenoid quinone structural types in bacteria and their taxonomic implication. *Microbiol. Rev.* **45**, 316–354.

Crabtree, K. T., McCoy, E., Boyle, W. C., and Rohlich, G. A. (1965). Isolation, identification, and metabolic role of the sudanophilic granules of *Zoogloea ramigera*. *Appl. Microbiol.* **13**, 218–226.

Doolittle, R. F. (1981). Similar amino acid sequences: Chance or common ancestry? *Science (Washington, D.C.)* **214**, 149–159.

Fahy, G. M. (1986). The relevance of cryoprotectant "toxicity" to cryobiology. *Cryobiology* **23**, 1–13.

Haynes, W. C., Wickerham, L. J., and Hesseltine, C. W. (1955). Maintenance of cultures of industrially important microorganisms. *Appl. Microbiol.* **3**, 361–368.

Jeffries, L. (1969). Menaquinones in the classification of Micrococcacea with observations on the application of lysozyme and novobiocin sensitivity test. *Int. J. Syst. Bacteriol.* **19**, 183–187.

Johnson, J. (1984). Bacterial classification III. Nucleic acid in bacterial classification. *In* "Bergey's Manual of Systematic Bacteriology" (P. H. A. Sneath, N. S. Mair, M. E. Sharpe, and J. G. Holt, eds.), Vol. 2, pp. 972–975. Williams & Wilkins, Baltimore, Maryland.

Jones, C. W. (1980). Cytochrome patterns in classification and identification including their relevance of the oxidase test. *In* "Microbial Classification and Identification" (M. Goodfellow and R. G. Board, eds.), pp. 127–138. Academic Press, New York.

Kandler, O. U., and Schleifer, K. H. (1980). Taxonomy I: Systematics of bacteria. *In* "Progress in Botany" (H. Ellenberg, K. Esser, K. Kubitzki, E. Schnepf, and H. Ziegler, eds.), Vol. 42, pp. 234–252. Springer-Verlag, Berlin and New York.

Kates, M. (1978). The phytanyl ether-linked polar lipids and isoprenoid neutral lipids of extremely halophilic bacteria. *Prog. Chem. Fats Other Lipids* **15**, 301–342.

Keddie, R. M., and Bousfield, I. J. (1980). Cell wall composition in the classification and identification of coryneform bacteria. *In* "Microbial Classification and Identification" (M. Goodfellow and R. G. Board, eds.), pp. 67–188. Academic Press, New York.

Kersters, K., and De Ley, J. (1975). Identification and grouping of bacteria by numerical analysis of their electrophoretic protein pattern. *J. Gen. Microbiol.* **87**, 333–342.

Krieg, N. R., and Holt, J. G., eds. (1984). "Bergey's Manual of Systematic Bacteriology," Vol. 1. Williams & Wilkins, Baltimore, Maryland.

Lapage, S. P., Shelton, J. E., Mitchell, T. G., and Mackenzie, A. R. (1970). Culture collections and the preservation of bacteria. *In* "Methods in Microbiology" (J. R. Norris and D. W. Ribbons, eds.), Vol. 3A, pp. 136–228. Academic Press, New York.

LaSalle, B. (1974). *In* "Round Table Conference on the Cryogenic Preservation of Cell Cultures" (A. P. Rinfret and B. LaSalle, eds.), pp. viii–xi. National Academy of Science, Washington, D. C.

Lechevalier, M. P. (1977). Lipids in bacterial taxonomy—A taxonomist's view. *Crit. Rev. Microbiol.* **5**, 109–20.

MacKenzie, A. P. (1977). Comparative studies on the freeze-drying survival of various bacteria: Gram type, suspending medium and freezing rate. *Dev. Biol. Stand.* **36**, 263–277.

Mazur, P. (1970). Cryobiology: The freezing of biological systems. *Science* (Washington, D.C.) **168**, 939–948.

Mazur, P. (1977). The role of intracellular freezing in the death of cells cooled at supraoptimal rates. *Cryobiology* **14**, 251–272.

Merchant, I. A. (1950). "Veterinary Bacteriology and Virology." Iowa State College Press, Ames, Iowa.

Meryman, H. T., ed. (1966). "Cryobiology." Academic Press, New York.

Meryman, H. T. (1971). Cryoprotective agents. *Cryobiology* **8,** 19–27.

Murray, R. G. E. (1984). The higher taxa, or, a place for everything . . .? *In* "Bergey's Manual of Systematic Bacteriology" (P. H. A. Sneath, N. S. Mair, M. E. Sharpe, and J. G. Holt, eds.), pp. 995–998, Williams & Wilkins, Baltimore, Maryland.

Rey, L.-R. (1977). Glimpses into the fundamental aspects of freeze-drying. *Dev. Biol. Stand.* **36,** 19–27.

Rowe, T. W. G. and Snowman, J. W. (1976). "Edwards Freeze-drying Handbook." Edwards High Vacuum, Inc., Grand Island, New York.

Sasser, M. (1990). Identification of bacteria by gas chromatography of cellular fatty acids. MIDI Tech. Note No. 101.

Selander, R. K., Cougant, D. A., Ochman, H., Musser, J. M., Gilmour, M. N., and Whittam, T. S. (1986). Methods of multilocus enzyme electrophoresis for bacterial population genetics and systematics. *Appl. Environ. Microbiol.* **51,** 873–884.

Shaw, N. (1975). Bacterial glycolipids and glycophospholipids. *Adv. Microb. Physiol.* **12,** 141–167.

Simione, F. P., and Brown, E. M., eds. (1991). "ATCC Preservation Methods: Freezing and Freeze-drying. American Type Culture Collection, Rockville, Maryland.

Sneath, P. H. A., and Sokal, R. R. (1973). "Numerical Taxonomy." W. H. Freeman and Company, San Francisco, California.

Sneath, P. H. A., Mair, N. S., Sharpe, M. E., and Holt, J. G., eds. (1986). "Bergey's Manual of Systematic Bacteriology," Vol. 2. Williams & Wilkins, Baltimore, Maryland.

Staley, J. T., and Krieg, N. R. (1984). Bacterial classification I. Classification of prokaryotic organisms: An overview. *In* "Bergey's Manual of Systematic Bacteriology" (P. H. A. Sneath, N. S. Mair, M. E. Sharpe, and J. G. Holt, eds.), Vol. 2, pp. 965–968. Williams & Wilkins, Baltimore, Maryland.

Stamp, L. (1947). The preservation of bacteria by drying. *J. Gen. Microbiol.* **I,** 251–265.

Tanaka, Y., Yoh, M., Takeda, Y., and Miwatami, T. (1979). Induction of mutation in *Escherichia coli* by freeze-drying. *Appl. Environ. Microbiol.* **37,** 369–372.

Yamada, Y., Inouye, G., Tahara, Y., and Kondo, K. (1976). The menaquinone system in the classification of coryneform and nocardioform bacteria and related organisms. *J. Gen. Appl. Microbiol.* **22,** 203–214.

CHAPTER 5

Actinomycetes

Alma Dietz
Sara A. Currie

Background

Introduction

Actinomycetes are microorganisms that share properties with both bacteria and fungi. They possess cell wall characteristics of the prokaryotes and morphological forms, such as filaments and conidia in chains, that are found in the eukaryotic fungi. The significance of morphology in preservation and maintenance of actinomycetes will be discussed in this chapter. Prior to the 1940s actinomycetes were of interest because of their distinctive colors and odors when cultivated on laboratory media, or because they were causative agents of disease in animals, humans, and plants. One of the actinomycetes pathogenic to humans is *Mycobacterium tuberculosis*. Actinomycetes are also the source of many compounds used to treat or prevent infectious diseases. From the mid-1940s to mid-1980s many pharmaceutical companies supported intensive isolation, fermentation, and characterization studies on actinomycetes in a search for products. Cultures producing valuable compounds had to be preserved so that the ability to produce the product of interest was not lost. Methods for preserving these valuable microorganisms are discussed in this chapter. There is also some discussion of characterization of actinomycetes. Individuals charged with preserving actinomycetes, like those concerned with other cell types, will need to authenticate the preserved material by verifying the retention of distinctive properties of the cultures.

MAINTAINING CULTURES FOR BIOTECHNOLOGY AND INDUSTRY
Copyright © 1996 by Academic Press, Inc.
All rights of reproduction in any form reserved.

Classification and Diversity

Actinomycetes occupy a unique niche in the scientific world. The funguslike filamentous growth of these organisms first led to publications in mycological journals (Hesseltine, 1960). Later studies of actinomycetes emphasized biochemical relationships that led to the differentiation of *Nocardia* and *Streptomyces* by paper chromatography of whole-cell hydrolysates (Becker *et al.*, 1964). Advances in microscopic and chemotaxonomic methods have greatly enhanced the abilities of scientists to differentiate genera of actinomycetes. Identification continues to remain difficult. Two references are helpful. One gives practical suggestions for examining actinomycetes (Cross, 1989); the other is a guide for identifying them to genus level (Lechevalier, 1989). The individual responsible for preserving actinomycetes must have a basic knowledge of the properties of actinomycete cultures. In 1948 three families and five genera of the Order Actinomycetales were recognized (Breed *et al.*, 1948). At the present time over 80 genera have been described (Dietz, 1988). Current classifications use suprageneric groups (Goodfellow, 1989). Other references (Buchanan and Gibbons, 1974; Kämpfer *et al.*, 1991; Sneath *et al.*, 1986; Williams *et al.*, 1989) are also useful for the taxonomist and the curator. It must be remembered that classifications reflect the thinking of the author(s), and that errors may occur, resulting in incorrect descriptions of a culture (Staley and Krieg, 1986).

Industrial Importance

In the mid-1940s pharmaceutical companies in the United States initiated screening programs to find and isolate cultures that could be induced in the laboratory to produce products for the treatment of infectious diseases. Following the discovery of actinomycin (Waksman and Woodruff, 1941) and streptomycin (Schatz *et al.*, 1944), it was suggested that soil should be used as a source of microorganisms with properties antagonistic to disease-producing organisms (Waksman and Woodruff, 1940; Waksman *et al.*, 1942). The interest in soils as a source of antibiotic-producing microorganisms resulted in 40 years of intensive competition by pharmaceutical industries to find such products. The search commenced in the United States but soon was undertaken by companies throughout the world. After World War II, the antibiotic streptomycin was credited with reducing the incidence of tuberculosis in Japan. Soon Japanese scientists were visiting the United States and learning how to develop screening programs. The Japanese interest in antibiotics led to the publication of the Journal of Antibiotics. A list of antibiotics and modifications of them can be found in a publication on antibiotic-producing *Streptomyces* (Queener and Day, 1986).

In addition to antibiotics, other metabolites of commercial value are produced by actinomycetes isolated from soil or obtained from culture collections [American Type Culture Collection (ATCC), 1991]. Products from cultures of industrial importance usually become the subject of a patent application, which contains a description of the producing organism(s). This description is an integral part of the application, and so is the deposit of the strain in a culture collection recognized by the U.S. Patent Office, the European Patent Office (EPO), and/or by the Budapest Treaty (Vossius, 1981). Deposits under the Budapest Treaty also meet the requirements of the U.S. Patent Office.

Characterization

The time from culture isolation to marketable product can easily exceed 10 years. The culture needed to produce the product must be described in a patent application used to protect the invention of a marketable item. In the 1960s taxonomists became frustrated because there were no standardized conditions published for use in cultivating and characterizing streptomycetes. At that time all actinomycetes isolated and found to produce antibiotics were thought to be *Streptomyces* spp. There was also concern that new species names were being assigned just to claim newness. This was a misconception on the part of those not responsible for determining the identity of the culture. There was no "handbook" on media formulations, growth conditions, morphological properties to observe, etc. A group of dedicated taxonomists organized and participated in the International *Streptomyces* Project (ISP), an international project to redescribe all cultures cited as *Streptomyces* available from public or private culture collections. The methods and media (Shirling and Gottlieb, 1966) are still useful and are recommended for characterization of *Streptomyces* spp. Difco Laboratories (Detroit, Michigan) supported the ISP by preparing ISP media in dehydrated form. Current availability of the ISP media should be determined by contacting your supplier. New descriptions were prepared for over 400 cultures examined in the cooperative study (Shirling and Gottlieb, 1968a,b, 1969, 1972). The ISP characterization publications remain the best set of descriptions of a group of actinomycetes. In addition to descriptions of cultures grown under standardized conditions, there are light microscopic and electron microscopic pictures of the sporulation structures of the cultures studied. A workshop manual (Dietz and Thayer, 1980) covers both the ISP methods and other methods for studying actinomycetes; these were not previously combined in one volume. The ability to detect an actinomycete on an isolation plate is a skill that must be developed through practical hands-on or "eyes-on" experience. Determining how to cultivate

the isolate is another required phase of the process. Determining the genus
and species is often not important to the industrial microbiologist until the
culture is found to make a product of interest. At that time macromorphol-
ogy (color, texture, pigment production, and odor) may be determined.
Then micromorphology must be determined. Distinctive spore chain morphology (Hunter-Cevera and Eveleigh, 1990)—for example, straight, open
spiral, tight spiral, fragmentation, sporangia, and motile spores can usually
be seen by light microscopy (LM). Confirmation of the presence of these
morphological features, and observations of spore chain development and
surface ornamentation of spores and other structures, are accomplished by
transmission and scanning electron microscopy (TEM and SEM). Chemotaxonomy is then used to confirm the generic status of the isolate. More
detailed culture observations are made using media known to be favorable
for cultivation of the suspected genus. The ISP media may be used for a
number of nonstreptomycete genera. Determination to the species level is
the next challenge. The culture may be considered similar to a known
species. Cultures assumed to be the same species must be grown for comparison under standardized conditions. Such cultures should have the same
color and microscopic characteristics. They will usually produce the same
antibiotic when fermentation conditions are identical. Suspected identity
may be supported by polyacrylamide gel electrophoresis (PAGE) (Dietz,
1988). Cultures of the same species will have the same characteristic protein
bands. Techniques developed in molecular biology laboratories that analyze
genotypic characteristics may also be used to compare cultures. DNA : DNA
hybridization studies and analysis of restriction fragment length polymorphism (RFLP) are two techniques coming into general use. However, these
useful procedures require special equipment and experienced personnel to
ensure valid identification of a culture (Park *et al.,* 1991). A listing of
actinomycete cultures that have been characterized and named is not readily
available to most taxonomists. To be accepted by the scientific community,
the description of a new culture must be validly published (Sneath, 1992).
The industrial actinomycete taxonomist and culture curator will be aware
of the publication of many new names and culture collection numbers in
the weekly patent publications covering different fields. The Derwent
weekly on pharmaceuticals is useful for those interested in antibiotics (Derwent Publications Limited, London, England). Characterization of cultures
in patents must be based on reproducible methods. Once the U.S. Patent
is issued, the culture is available to the public. Anyone obtaining and
cultivating the culture should be able to reproduce the results reported in
the patent. If this is not possible, the patent may be invalidated. Keeping
track of culture descriptions in patents and other publications is important
to the person responsible for characterizing actinomycete cultures for an

industrial laboratory. The culture being investigated may be the same as one described in a competitor's patent. There are many abstract and journal reference sources that should be checked regularly for new names and culture collection numbers. A useful personal reference file and/or an in-house computer search system can be developed. An image database for identification of actinomycetes (Ugawa *et al.,* 1989) combines descriptions of cultures with a visual display of morphology observed with the scanning electron microscope. It is suggested that the curator of actinomycetes read the short review (Seino, 1991) of actinomycete culture collections.

Methods for Preservation

Methods for handling actinomycetes vary. One person may report a method worthless while another will report it to be the only one to use. Why are there such strong differences of opinion? It will soon be obvious that not all steps used in preparing actinomycetes for preservation are the same in every laboratory. To report that "lyophilization" or "soil" or "liquid nitrogen" was used does not truly describe the method employed. How was the culture cultivated? On what medium was it grown? Was distilled, deionized, or tap water used? What incubation temperature was used? How long was the culture incubated? If suspensions were stored in soil or made for lyophiles, how long did the culture stay in the suspension medium? Was it used in the log phase or in the lag phase? What was the suspension medium? How was the preparation dehydrated for lyophile preparation? What type of soil was used? How were the sterile soil tubes prepared? If liquid nitrogen was used, was the culture stored in the liquid phase or in the gas phase?

All of these questions are important in deciding which preservation method(s) may be best for a strain. Restoration conditions (medium, temperature, air, tube, plate, or shake flask cultivation) affect the viability of the preserved culture and the value of the preservation method(s) used. The importance of the restoration conditions is often overlooked; the conditions used to restore the culture may differ from those used for the cultivation of a culture for a specific process.

Culture Condition

The physical appearance of the actinomycete to be preserved should be observed and recorded. One way to do this is to take color photographs of the culture on several diagnostic media. These should be supplemented with LM and TEM or SEM photographs of microscopic properties. Comments on observed properties can be entered into a computer search program. The curator then has a useful reference profile of a culture grown under standardized conditions. In an in-house situation and in an unrestricted outside deposit, the recipient of a culture may demand that certain criteria be met. This is not always possible with a restricted patent deposit. However, any problems with such cultures will be detected as soon as the cultures become publicly available.

The culture must grow well and exhibit the specific properties cited for it in a patent or a scientific publication. It must be free of microbial contaminants. It is assumed that most actinomycetes carry an actinophage. This is usually detected as plaques on the surface of cultures grown on agar slants or plates, and a clearing or thinning of broth cultures. A lysogenic actinophage can wreck havoc. An expensive antibiotic fermentation may be destroyed (Saudek and Collingsworth, 1947; Carvajal, 1953). Soil isolates and cultivated collection strains may harbor obscure contaminants. A supposedly pure soil isolate may actually be two isolates; one may flourish at 24–28°C, the other at 37–45°C. This problem would go undetected until temperature studies were performed. Fungal spores may become airborne and infect a culture being transferred. The fungus would not be detected until the actinomycete was cultivated on a medium favorable to the fungus as well as to the actinomycete. Microbial contaminants may be detected by microscopic observation and cultivation of the culture on media selective for a suspected contaminant.

GROWTH CONDITIONS
(pH, MEDIA, TEMPERATURE)

The optimum growth conditions must be determined for each actinomycete being preserved. Just because the culture is a *Streptomyces* or a *Micromonospora,* for example, does not mean that all *Streptomyces* spp. will exhibit optimum growth on the medium recommended for one species. Does the culture require a high or a low pH? Does it grow better and produce spores in a synthetic or a natural products medium? What is its optimum growth temperature? What is its temperature range? Is it thermoduric or thermophilic? For example, most blue-spored, melanin-positive streptomycetes will grow well in 24 hours at 45°C.

PHYSIOLOGICAL STATE (TITER, DILUENT,
GROWTH PHASE, SPORULATION)

Actinomycetes are prolific producers of unusual metabolites, especially antibiotics. For shake flask fermentations, media conditions must be found that will stimulate cultures to produce one or more metabolites extracellularly, intracellularly, or both. The curator of actinomycetes must be aware of how a given culture is induced to elaborate a metabolite. Strain development, selection, and fermentation development are needed to obtain a high-yielding culture. The strain is then propagated under rigid conditions (standardized media formulation and autoclave time). It must be maintained for future fermentations and/or until an even better strain (yielding a higher titer) is obtained. Properties of a culture (morphology) and effects of medium pH, medium ingredients, temperature, aeration, and rate of agitation of the culture in various sizes of shaken flasks or large fermentation vessels or tanks affect the final titer (Bader, 1986). Actinomycete cultures are spore formers. They may produce spores both aerially and in the submerged state (on solid media and in shaken liquid fermentation media). How and under what conditions spores are formed must be determined for the culture under study.

Continuous Subculture

Continuous subculture is one way to maintain cultures. It may be the only way to do so in situations wherein refrigeration, lyophilization equipment, and personnel are in short supply or are nonexistent. In some situations this may be a costly use of manpower. It may lead to loss of the culture. Degeneration can occur rapidly if a culture is maintained on the same nutrients. Alternation of growth media is essential to reduce this effect. The culture may be grown on a rich medium; at transfer time it may be subcultured to a minimal or synthetic medium. On the next transfer it is subcultured to a different rich medium. Transfer time is also variable. For example, stock cultures maintained on agar slants may be transferred annually as a protective or preventive loss measure in facilities with no other means of storage. Cultures of certain production strains may be transferred semiweekly or monthly to keep adequate stocks available. Culture rundown is also a problem in continuous fermentations (Bader, 1986; Reusser et al., 1961). Subculture to agar slants or agar plates is done much as it is with fungi, rather than with bacteria. Fungi form hard mats of aerial growth; unicellular bacteria form soft colonies. The filamentous actinomycetes form hard colonies or mats on an agar surface. A firm (nichrome) wire loop, or its equivalent, should be used to "pick up" or dislodge the growth. This type of loop also works well for the soft coryneform and nocardiaform colonies, which have a superficial resemblance to the unicellular bacteria. A loopful of growth is transferred to the fresh agar and is spread with the loop. A more uniform growth will occur if the growth to be transferred is suspended in sterile distilled water, again using the sterile nichrome loop. Inoculum (0.2 ml) is then pipetted onto the agar surface. On plates the inoculum is spread in a cross-hatch. The curator can determine if a contaminant is present by viewing the intersections with the microscope. Depending on the use of the culture, it can be streaked down the center of a slant or spread uniformly over the surface. Uniform spreading will yield more material for use in culture inoculum preparation.

Soil Preparations

Preservation in sterile sandy loam soil may be the most practical and cost-efficient way to preserve filamentous spore-forming microorganisms. Untreated soil (from an organic garden or farm or a good wooded site that has not been fertilized or treated with pesticides) should be used. The soil should be distributed in a shallow layer in a metal tray and heated at least 6 hours at 150°C to destroy spore-forming microorganisms or insects that may be present. After the soil has cooled, any debris is removed. The soil is then sieved with 10- and 20-mesh sieves. If finer soil is desired it may then be passed through a 30-mesh sieve. Store the soil in a dry place in covered containers. Test tubes of soil may be prepared to suit the needs of the user. A convenient size is Pyrex (Trademark Reg.) 13 × 100 mm. Dispense the dried and sieved soil to a depth of 20 mm. Plug the tubes with gauze-covered nonabsorbent cotton. Autoclave for 60 min on slow exhaust on two successive days. On the third day autoclave for 60 min or fast exhaust and dry. Store sterile tubes with a loose cover. Confirm sterility by seeding a medium supporting bacterial growth with a loopful of soil from a tube in the middle of the rack of sterile soil tubes. Incubation at 24, 28, and 37°C is recommended for the sterility check. A sterile distilled water suspension of growth from a well-sporulated culture on an agar slant gives very good results. Use 1 ml of suspension to saturate soil in a tube. Sterile distilled water, used as the suspending agent, eliminates problems that may occur if broth is used. [If broth is used, the culture will grow in the broth and produce self-toxic metabolites. This may be the reason for problems reported with actinomycetes stored in soil (Pridham *et al.,* 1973).] Air-dry the wet soil tube at room temperature (24°C) for about a month, tap the tube on the laboratory bench (disinfected surface) to loosen the soil particles, and store at 4°C in a temperature-controlled refrigerator. To obtain inoculum from the stored soil tube use a sterile inoculating loop moistened in sterile distilled water. The soil particles will be held on the wet loop. It is then easy to streak the inoculum on a solid medium of choice or inoculate a broth medium.

Dried Blood Preparations

There are reports in the literature that actinomycete numbers increase when dried blood is added to soil samples used for isolation of these organisms (Waksman and Starkey, 1924; Porter and Wilhelm, 1961). Thus it may be assumed that addition of sterile dried animal blood to a suspension of the organism will enhance or stabilize the culture being stored. Mary P. Lechevalier (Waksman Institute of Microbiology, Rutgers, The State University of New Jersey) recommends the following medium:

Soil, potting or rich organic	100.0 g
CaCO$_3$	10.0 g
Dried blood	2.5 g

The dried blood may be replaced with blood fibrin and hemoglobin (10 : 1), if desired. Sterilize the soil 1 hour at 15 lb. Incubate at 37°C, overnight. Repeat this twice. Add the dry blood and CaCO$_3$, mix well, and resterilize. Check sterility by inoculating some soil mixture into a rich broth medium. Add 20 ml of sterile water, shake slightly, and inoculate with a spore suspension of the actinomycete. Incubate for 2–5 weeks.

Lyophilization

Lyophilization is perhaps the method most widely used for long-term storage of microorganisms. Liquid is removed from suspensions of the frozen cells under reduced pressure and the dried cells may be maintained for an indefinite time period.

To protect cells from ice crystal damage during the freezing portion of the lyophilization procedure, a cryoprotective agent, such as sterile 15% noninstant skim milk solution, is used as the suspending agent. A 10- to 14-day well-sporulated slant is scraped and aliquots are placed in borosilicate glass ampoules with sterile cotton plugs. These ampoules are then frozen in a dry ice–solvent bath.

A freezer-dryer is prepared according to the manufacturer's directions. The shell temperature should be at −40°C and the condenser at −50°C. As soon as the frozen ampoules are placed inside the chamber, the vacuum pump is started. When the vacuum is below 100 μm Hg, the shell temperature can be raised to 10°C and the lyophilizer is run until the culture suspension has dried, usually overnight.

When dried, the ampoules are attached to a vacuum manifold and are sealed under vacuum using an oxygen-gas torch. After sealing, all ampoules are tested for vacuum using a Tesla coil. Ampoules with weak or no vacuum are discarded because long-time viability is affected. Ampoules should be stored in the dark, preferably at 4°C.

To revive the culture, the ampoule is scored, the outer surface is sterilized, and the ampoule is broken open while contained in a sterile gauze pad, minimizing the aerosol effect. The pellet is dropped into 2 ml of an appropriate growth medium.

There are many procedures for lyophilizing (freeze-drying) microorganisms. Also, there are many types of equipment, homemade and commercial, that may be found suitable for any laboratory. Procedures for the preservation of *Streptomyces* spp. by lyophilization are described in two manuals dealing with actinomycetes (see Hopwood *et al.,* 1985; Pridham and Lyons, 1981). For a better understanding of lyophilization, refer to Gherna (1981).

Cryopreservation (Straw)

A method that is rapid and reliable is the cryopreservation of agar plugs of cultures in the gas phase of liquid nitrogen. Well-sporulated, poorly sporulated, and asporogenous cultures are grown on an agar medium in petri plates. Cultures are grown for 10–14 days at their optimum growth temperature. The growth is then plugged into straws that have been sterilized in screw-cap vials (Dietz, 1974). Agar acts as a cryopreservative. The vials containing the straws are stored in the gas phase of liquid nitrogen. About five plugs are in each straw. One plug may be removed aseptically from a straw and the straw returned to its storage container. Variations on this may be introduced as necessary. Suspensions of spores and vegetative cells, with or without the addition of a cryopreservative such as glycerol, may be dispensed in ampoules. The ampoules can be sealed and stored under liquid nitrogen (liquid phase). Suspensions may also be placed in screw-cap vials and stored in the gas phase of liquid nitrogen. These preparations and the agar-plug preparations may also be stored in low-temperature ($-70°C$) mechanical refrigerators. Under this type of storage the culture is exposed to air and may grow, although at a slow rate. It is believed that for long-term storage the liquid nitrogen storage is preferable because of the lack of oxygen, which can support growth.

Cryopreservation (Vials)

In preserving cultures by freezing, a cryoprotective agent must be used to protect the cells from damage. The suspending medium in most common use is 10% glycerol. DMSO can also be used in cases in which its skin-penetrating property does not present a safety hazard.

The 10- to 14-day well-sporulated slants can be scraped with sterile 10% glycerol and aliquots of the suspension placed in sterile borosilicate glass vials with Teflon-lined screw caps. Alternatively, a 3-day shake flask culture broth can be mixed with an equal volume of sterile 20% glycerol and aliquots prepared.

The vials can then be frozen in the gas vapor phase of a liquid nitrogen freezer or in a $-80°C$ mechanical freezer. Vials should not be stored in the freezer unit of self-defrosting home-style refrigerators. The $-20°C$ temperature of these units is in the range where the most cell damage occurs. The heating portion of the defrost cycle also causes a small thawing and refreezing effect that can result in reduced survival of cells. If sterile Nunc vials are used and placed in a plastic sleeve, the vials can be stored directly in liquid nitrogen. However, this storage requires special handling and is necessary only when the $-196°C$ temperature of liquid nitrogen is required for survival.

To revive the culture, rapid thawing at $37°C$ and immediate use provide the best results.

After a batch has been prepared and frozen for at least 72 hours, vials should be opened and the culture tested for purity, viability, stability of phenotypic characteristics, and retention of any special properties (antibiotic production, resistant markers, etc.).

Other Methods for Actinomycete Preservation

Many different methods can be used to preserve microorganisms (Dietz, 1981). One of these methods involves drying inoculum on sterile glass beads in vials and storing the vials in the gas phase of liquid nitrogen or in a mechanical refrigerator. Inoculum may also be dried on sterile $\frac{1}{4}$- or $\frac{1}{2}$-inch assay discs. The dried discs may be frozen or stored in sterile vials in the refrigerator. Cultures growing on agar slants may be frozen ($-70°C$ deep freeze) and aliquots chipped off as needed. Not all methods used to preserve actinomycetes are found in the literature. It is the responsibility of the curator to find a reliable, reproducible method for preservation of the cultures that are his or her responsibility. Often this may involve modifications of described methods.

Evaluation of Viability and Stability

In every discussion of culture preservation the significance of viability and stability is emphasized. Much time and expense can be incurred in determining the optimum conditions for some cultures. Can this be justified? Yes, if the culture is important to a manufacturing process (Dietz and Churchill, 1986). It may be of little concern to a teaching laboratory that can easily obtain a new strain from an inexpensive source.

The industrial actinomycete taxonomist should be very concerned about the suitability of preserved actinomycete cultures. These strains are essential to comparative studies that must be made when a new strain is important to a patent application. The well-preserved culture is important also to research studies. Restored cultures must have the properties exhibited before preservation. They must grow well in comparative studies using standardized conditions. Specifically, actinomycetes might be grown on ISP media 2–5 (Shirling and Gottlieb, 1966) in four-sector plates. Most actinomycetes give a good color pattern and produce reproductive structures necessary for microscopic characterization. Characteristics observed can easily be compared with those observed prior to storage. Growth on

ISP media 6 and 7 is useful for determining melanin production; the recommended incubation temperature must be used. In an industrial laboratory, production of the product of interest can be confirmed by a fermentation specialist; the presence of plasmids can be confirmed by a molecular biologist. These confirmation tests are time-consuming and may not be done on a routine basis because of lack of technical assistance. The use of polyacrylamide gel electrophoresis has become an essential taxonomic tool for species confirmation by comparison of protein bands. Thus, the ability to confirm the presence of plasmids may also become a routine operation in the culture collection laboratory. The plasmid stability of strains used by the molecular biologist must also be considered. How stable are the cultures with naturally occurring and/or inserted plasmids? In summary, the collection of preserved actinomycetes must be supported by quality control and information resource materials.

Selection of a Preservation Procedure

Choosing the best procedure for preserving cultures depends on many factors: the genera being preserved, the special properties of the culture, the number of cultures preserved as part of the normal work load, the equipment and storage areas available, the number of personnel assigned to the lab and their degree of experience, and the needs of the research or production areas served.

One method may be selected for routine work. Evaluation of each culture after preservation will alert personnel that a particular method may not be successful for a culture, and another preservation method should be tried. However, the effectiveness of any preservation procedure is determined not only by survival rate but also by retention of any special properties of interest. A study by Monaghan and Currie (1985) shows that a preservation method that is best for viability is not necessarily the best for retention of special properties. It may be necessary to accept a somewhat lower survival rate in order to maintain the special uses. Knowledge of the reason for preserving the culture, awareness of various preservation methods, and evaluation of preserved material should enable personnel to find a method that successfully preserves the culture.

General Concerns Relevant to All Methods

Several important factors must be considered for any preservation method.

BATCH TESTING

Samples of each preservation batch must be tested for viability, purity, stability of phenotypic characteristics, and retention of special properties before the culture is distributed for use. Over a long storage period special cultures should be tested at stated intervals.

DOCUMENTATION

Proper documentation is necessary for each batch prepared. This record should include the source of and kind of culture material being preserved, brief descriptions of the culture before and after preservation, results of viability and use tests, procedure used, date, number prepared, where stored, and personnel preparing the batch.

SAFETY

Well-maintained biological hoods, pipetting devices, proper equipment for handling frozen material, and training of personnel are important in compiling a good safety record.

SECURITY

When a culture is necessary to the economic health of a company, it is important to prevent complete loss of that culture to a research program or to a special collection. A few samples of preserved material should be stored away from the major storage area. All mechanical and liquid nitrogen freezers should be equipped with alarms and a procedure set up to alert concerned personnel that malfunction of freezers has occurred. Limited access to culture storage areas and documentation of personnel receiving preserved cultures will prevent unauthorized use or unnecessary use that depletes inventory. This is especially necessary for cultures that are difficult to preserve.

Acknowledgments

The authors thank their former employers, Merck & Co., Inc., and The Upjohn Co., for the opportunity to study and work with preservation of actinomycetes.

References

American Type Culture Collection (ATCC) (1991). "Microbes and Cells at Work," 2nd ed. ATCC, Rockville, MD.
Bader, F. G. (1986). In "The Bacteria" (S. W. Queener and L. E. Day, eds.), Vol. 9, pp. 281–321. Academic Press, Orlando, FL.
Becker, B., Lechevalier, M. P., Gordon, R. E., and Lechevalier, H. A. (1964). Appl. Microbiol. 12, 21–423.
Breed, R. S., Murray, E. G. D., and Hitchens, A. P., eds. (1948). "Bergey's Manual of Determinative Bacteriology," 6th ed. Williams & Wilkins, Baltimore.

Buchanan, R. E., and Gibbons, N. E., eds. (1974). "Bergey's Manual of Determinative Bacteriology," 8th ed. Williams & Wilkins, Baltimore.

Carvajal, F. (1953). *Mycologia* **45,** 209–234.

Cross, T. (1989). *In* "Bergey's Manual of Systematic Bacteriology" (S. T. Williams, M. E. Sharpe, and J. G. Holt, eds.), pp. 2340–2343. Williams & Wilkins, Baltimore.

Dietz, A. (1974). *In* "Round Table Conference on the Cryogenic Preservation of Cell Cultures" (A. P. Rinfret and B. LaSalle, eds.), pp. 22–36. National Academy of Sciences, Washington, DC.

Dietz, A. (1981). *In* "Biotechnology 1" (H.-J. Rehm and G. Reed, eds.), pp. 411–434. Verlag Chemie, Weinheim, Germany.

Dietz, A. (1988). *In* "Biology of Actinomycetes" (Y. Okami, T. Beppu, and H. Ogawara, eds.), pp. 203–209. Jpn. Sci. Soc. Press, Tokyo.

Dietz, A., and Churchill, B. W. (1986). *In* "Comprehensive Biotechnology" (M. Moo-Young, C. L. Cooney, and A. E. Humphrey, eds.), pp. 37–49. Pergamon, New York.

Dietz, A., and Thayer, D. W., eds. (1980). "Actinomycete Taxonomy" SIM Spec. Publ. No. 6. Soc. Ind. Microbiol., Arlington, VA.

Gherna, R. L. (1981). *In* "Manual of Methods for General Bacteriology" (P. Gerhardt, ed.), pp. 208–217. Am. Soc. Microbiol., Washington, DC.

Goodfellow, M. (1989). *In* "Bergey's Manual of Systematic Bacteriology" (S. T. Williams, M. E. Sharpe, and J. G. Holt, eds.), pp. 2333–2339. Williams & Wilkins, Baltimore.

Hesseltine, C. W. (1960). *Mycologia* **52,** 460–474.

Hopwood, D. A., Bibb, M. J., Chater, K. F., Kieser, T., Bruton, C. J., Kieser, H. M., Lydiate, D. L., Smith, C. P., Ward, J. M., and Schrempf, H., eds. (1985). "Genetic Manipulation of Streptomyces, A Laboratory Manual." pp. 356, The John Innes Foundation, Norwich, England.

Hunter-Cevera, J. C., and Eveleigh, D. E. (1990). *In* "Soil Biology Guide" (D. L. Dindal, ed.), pp. 33–47. Wiley, New York.

Kämpfer, P., Kroppenstedt, R. M., and Dott, W. (1991). *J. Gen. Microbiol.* **137,** 1831–1891.

Lechevalier, H. A. (1989). *In* "Bergey's Manual of Systematic Bacteriology" (S. T. Williams, M. E. Sharpe, and J. G. Holt, eds.), Vol. 4, pp. 2344–2347. Williams & Wilkins, Baltimore.

Monaghan, R. L., and Currie, S. A. (1985). *Dev. Ind. Microbiol.* **26,** 787–792.

Park, Y.-H., Yim, D.-G., Kim, E., Kho, Y.-H., Mheen, T.-I., Lonsdale, J., and Goodfellow, M. (1991). *J. Gen. Microbiol.* **137,** 2265–2269.

Porter, J. N., and Wilhelm, J. J. (1961). *Dev. Ind. Microbiol.* **2,** 253–258.

Pridham, T. G., and Lyons, A. J. (1980). *In* "Actinomycete Taxonomy" (A. Dietz and D. W. Thayer, eds.), SIM Spec. Publ. No. 6, pp. 152–254. Soc. Ind. Microbiol., Arlington, VA.

Pridham, T. G., Lyons, A. J., and Phrompatima, B. (1973). *Appl. Microbiol.* **26,** 441–442.

Queener, S. Q., and Day, L. E., eds. (1986). "The Bacteria," Vol. 9. Academic Press, Orlando, FL.

Reusser, F., Koepsell, H. J., and Savage, G. M. (1961). *Appl. Microbiol.* **9,** 342–345.

Saudek, E. C., and Collingsworth, D. R. (1947). *J. Bacteriol.* **54,** 41–42.

Schatz, A., Bugie, E., and Waksman, S. A. (1944). *Proc. Soc. Exp. Biol. Med.* **55,** 66–69.

Seino, A. (1991). *Actinomycetologica* **5,** 72–77.

Shirling, E. B., and Gottlieb, D. (1966). *Int. J. Syst. Bacteriol.* **16,** 313–340.

Shirling, E. B., and Gottlieb, D. (1968a). *Int. J. Syst. Bacteriol.* **18,** 69–189.

Shirling, E. B., and Gottlieb, D. (1968b). *Int. J. Syst. Bacteriol.* **18,** 279–399.

Shirling, E. B., and Gottlieb, D. (1969). *Int. J. Syst. Bacteriol.* **19,** 391–512.

Shirling, E. B., and Gottlieb, D. (1972). *Int. J. Syst. Bacteriol.* **22,** 265–394.

Sneath, P. H. A. (1992). "International Code of Nomenclature of Bacteria," 1992 Revision. Am. Soc. Microbiol., Washington, D.C.

Sneath, P. H. A., Mair, N. S., Sharpe, M. E., and Holt, J. G., eds. (1986). "Bergey's Manual of Systematic Bacteriology," Vol. 2. Williams & Wilkins, Baltimore.

Staley, J. T., and Krieg, N. R. (1986). *In* "Bergey's Manual of Systematic Bacteriology" (P. H. A. Sneath, N. S. Mair, M. E. Sharpe, and J. G. Holt, eds.), Vol. 2, pp. 965–968. Williams & Wilkins, Baltimore.

Ugawa, Y., Sugawa, K., Kudo, T., *et al.* (1989). *In* "Trends in Actinomycetology in Japan" (Y. Koyama, ed.), pp. 17–19. Society for Actinomycetes, Japan.

Vossius, V. (1981). *Bio Technology* **1,** 435–452.

Waksman, S. A., and Starkey, R. L. (1924). *Soil Sci.* **17,** 373–378.

Waksman, S. A., and Woodruff, H. B. (1940). *J. Bacteriol.* **40,** 581–600.

Waksman, S. A., and Woodruff, H. B. (1941). *J. Bacteriol.* **42,** 231–249.

Waksman, S. A., Horning, E. S., Welsch, M., and Woodruff, H. B. (1942). *Soil Sci.* **54,** 281–296.

Williams, S. T., Sharpe, M. E., and Holt, J. G., eds. (1989). "Bergey's Manual of Systematic Bacteriology," Vol. 4. Williams & Wilkins, Baltimore.

CHAPTER 6

Fungi

David Smith
Jacqueline Kolkowski

Background

Classification and Diversity

The fungi comprise a vast and varied group of organisms. There are at present over 70,000 described and accepted species; additionally, the unexplored areas of the world undoubtedly contain many undiscovered species. The most recent estimate of their total number is now conservatively considered to be 1.5 million (Hawksworth, 1991), about six times that of vascular plants. Around 1000 new species of fungi are recognized each year, i.e., at a rate equivalent to discovering a third of all accepted species of bacteria every year.

Fungi have the ability to exploit most ecological niches and the ability to utilize most natural and man-made products. Their capacity to change and adapt makes them ideal pioneers. They are found growing at low temperature in Antarctica, where they grow on and in rocks at temperatures of -10 to $-20°C$, withstanding temperatures as low as -70 to $-80°C$ in severe conditions. The genus *Aureobasidium* contains species that withstand such extreme temperatures and also includes representatives that grow in temperate and tropical regions. The vast majority of fungi are aerobic and require high water activity to grow; however, anaerobic fungi have been discovered in the ovine rumen (Yarlett *et al.*, 1986), and many are capable of growing in substrates with low water activity, such as jam. The hosts

MAINTAINING CULTURES FOR BIOTECHNOLOGY AND INDUSTRY

101

index in the *Index of Fungi* cumulative index for 1981–1990 (International Mycological Institute, 1991) illustrates the vast array of hosts and substrates from which fungi are isolated. They are found on, associated with, or causing disease of algae, bryophytes, vascular plants, other fungi (including lichens), arthropods, other invertebrates, and vertebrates. Substrates such as air, soil, water, dung, food, plant products, petroleum, and pharmaceutical products are also listed as sources of fungi.

Their classification is inevitably complex, and is generally based on anatomical and morphological characters (Table 1). The crucial characteristics that are used to classify the main fungal groups include the development and structure of specialized bodies containing the spore-producing organs, the precise nature of the organs themselves, the nature of the cell walls, and the occurrence of and the types of flagellae. Other characteristics become important at the order, family, genus, and species levels, for example, details of fruiting body structure and development, spore color, septation, ornamentation, and size. The vast literature on the taxonomy of different groups of fungi reflects their large numbers. This also makes the identification of an organism quite critical when it is to be used as a reference, research, or production strain. Misidentified organisms could invalidate research conclusions, and can waste valuable resources. The yeasts are normally studied separately from the filamentous fungi. However, they are growth/habit forms produced by a wide range of different fungal groups and many have characteristics of the Endomycota and the Ustomycota in particular (see Table 1). Some yeasts have filamentous states; for example, species of the genus *Schizosaccharomyces* may produce true hyphae and arthrospores (Kreger-van Rij, 1984). The reverse is also true where some filamentous fungi have yeastlike states, for example, *Amylomyces rouxii.* [For access to the mycology literature see Barnett *et al.* (1983), Gams *et al.* (1987), Hawksworth *et al.* (1983), Hawksworth and Kirsop (1988), Kirsop and Kurtzman (1988), Kreger-van Rij (1984), Lodder and Kreger-van Rij (1952), Onions *et al.* (1981), and Samson and van Reenen-Hoekstra (1988).] The *Bibliography of Systematic Mycology* of the International Mycological Institute (IMI) provides an ongoing source of systematic and identification literature, from the specialist down to the more general level.

The identification of individuals from such a vast group of organisms requires specialist knowledge. Although it is possible to get access to relevant literature, the expertise necessary to name many organisms correctly comes from examining many representatives of a taxonomic group. Culture collections must be accessible to specialists to identify and check the quality of the strains they collect and supply. Taxonomic links are quite often used when selecting strains for screening for particular properties. It is therefore crucial that the strains are correctly named and that specialist advice is

TABLE 1
Some Characteristics Used in the Classification of Fungi[a]

Kingdom	Phyla	Characteristics
Protoctista	Myxomycota	Amoeboid; many form aggregations, plasmodia, or sporophores
	Chytridiomycota	Not amoeboid; many unicellular; chitinoid walls; flagellate zoospores
Chromista	Oomycota	Filamentous; cellulose wall; motile zoospores with two types of flagellae; sexual state oogonia
Fungi	Zygomycota	Filamentous; produce zygospores (resting spore) as sexual state
	Endomycota[b]	Endospores produced inside cells; lacking special fruit bodies; mostly yeastlike; septa, where present, not centrally perforate; chitin often absent
	Ascomycota[b]	Ascospores found inside an ascus; usually in specialized fruit bodies; most are filamentous; hyphal septa centers perforate; walls not multilayered; electron transparent; chitin present
	Basidiomycota[b]	Basidiospores produced outside, usually nonseptate basidia; clamp connections and dolipore septa usually present; multilayered and electron-dense walls; most are filamentous
	Ustomycota[b]	Basidiospores produced outside, cross-septate basidia; with clamp connections; dolipore septa absent; spores germinate to produce a yeast like phase that can predominate; walls lacking xylose

[a] Modified from Hawksworth (1991).
[b] *Note:* Members of these phyla, in which sexual (teleomorphic) states are generally absent, are sometimes called deuteromycetes, "conidial," or "imperfect"; most are anamorphs of ascomycetes.

sought when tracing such links. This is also true for organisms to be used as reference, test, or assay strains. In many cases a specialist's advice will be required to select the most suitable strain for a particular purpose.

Industrial Importance

Fungi are utilized by humans in many ways and the uses are expanding rapidly. The use of large fungi or mushrooms as food is quite commonplace, but more recently the hyphae of *Fusarium* have been produced in such a way that they can be sold as a meat substitute, Quorn mycoprotein (Trinci, 1992). The high nutrient content and high level of marketing are making this product quite successful. Mold-ripened cheese is another major use of fungi; *Pencillium roquefortii* is able to grow in the narrow spaces of the curd, with a small amount of oxygen, to the near exclusion of all other molds.

Enzymes are also produced by fungi commercially: *Aspergillus oryzae* and *Aspergillus niger* are sources of α-amylase, and *Trichoderma* species produce exploitable levels of cellulases. Fungi are also used to detoxify waste by concentrating heavy metals and to compost organic wastes.

Fungi are also a very important source of industrial chemicals, for example, citric acid and gluconic acid from *Aspergillus niger* and itaconic acid from *Aspergillus terreus*. Alcohol produced by fermentation from *Saccharomyces* species is a well-known, large industry.

There is increased interest in screening for natural products important for human health and well being, and the traditional use of fungi as antibiotic producers has been extended to the use of their products as antitumor agents and immunosuppressants. Griseofulvin is an antifungal agent produced by *Pencillium griseofulvum* and the immunosuppressant cyclosporin is a product of *Tolypocladium inflatum*.

The list of uses is extensive and although the long-term preservation of such process cultures is essential, it is equally important that their crucial properties are also retained for them to remain viable. In agriculture, insect pathogens such as *Beauveria bassiana* and *Metarhizium anisopliae* are used as biocontrol agents and rust fungi are being utilized for the control of weeds.

Characterization

Characterization of a strain is important for taxonomic purposes, for patent purposes, and to increase the value of strains to culture collections and their clients. The great diversity of the fungi extends far beyond morphological differences. The complexity of fungal metabolism and ability to adapt and change is reflected in the characters they express. Their potential is enormous. Biochemical and physiological activities have been increasingly used in attempts to define both formal taxonomic groups and discrete populations within species (Bridge and Hawksworth, 1990). The use of thin-layer chromotography (TLC) has generated a large amount of data on fungal metabolites (Paterson, 1986) and this has been added to profiles developed by high-performance liquid chromatography (Frisvad, 1989). Electrophoretic techniques to examine isoenzyme patterns and DNA hybridization have also been used to characterize fungi. The combination of morphological characters, physiology, and biochemical analyses of fungi has led to an integrated approach to aid in identifying them. Such systems reveal useful information that can be utilized in biotechnology (Bridge, 1985).

Fungi have also been characterized by growth in the presence of inhibitors and at different temperatures and pH values, by assimilation of carbon and nitrogen sources, and by enzyme activities (Bridge, 1985). The APIZYM testing system (Biomerieux, France) has been used as a rapid method to characterize fungi and can give additional useful information (Bridge and Hawksworth, 1984). The review of these and other techniques compiled by Bridge and Hawksworth (1990) covers key aspects of the chemotaxonomy of *Aspergillus, Beauveria, Fusarium, Mucor, Penicillium,* and *Rhizopus* species.

Methods for Preservation

Fungi can be difficult to maintain in a viable, pure, and stable condition. It is essential that appropriate growth conditions and preservation techniques are used to ensure stability of microorganisms for research and applications. With the development of biotechnology and bioengineering, it is becoming increasingly important to retain all characteristics throughout storage.

Many filamentous fungi grow on culture media and can be kept viable by periodic transfer. However, the properties of fungi in culture may be unstable through loss of plasmids, spontaneous mutations, or genetic recombination due to the presence of heterokaryons, the parasexual cycle, or normal sexual events (Smith, 1986). These phenomena can result in modification of a strain's characteristics. Therefore conditions of storage should be selected to minimize the risk of such changes. Culture preservation techniques may include continuous growth, reduction of metabolism, and halted metabolism. It is not known if the latter situation is achieved, but metabolism is reduced, by some techniques, to such a level that it is considered as suspended or nearly so (Smith and Onions, 1983a).

No one technique has been successfully applied to all fungi (Table 2), although storage in liquid nitrogen appears to approach the ideal. Most fungi that grow well in culture survive freezing in liquid nitrogen. Those that grow poorly tend to do less well, but even nonsporulating fungi can survive. Also, liquid nitrogen storage is a method that enables the preservation of fungi that do not grow in culture. For example, fungal pathogens in infected tissue can be preserved, bearing in mind that a mixed population of organisms may be present. Centrifugal freeze-drying allows only the more robust spores to survive (Smith, 1983a). Some sclerotia and other resting stages, and in a few cases sterile mycelia, have been known to survive freeze-drying (Smith and Onions, 1983a). Of the other techniques described in this chapter, storage in mineral oil, soil, or water is of use for a wide range of organisms. Studies at IMI (Smith and Onions, 1983a) have shown that the majority will survive, but the period of storage for some may be very short (Table 3). Only sporulating fungi survive storage in silica gel, and spores with thin walls and high water content or those with appendages do less well. The use of the best method available is essential for the preservation of irreplaceable isolates and for the maintenance of genetic stability. Oomycete fungi belonging to the Mastigomycotina are very vulnerable to harsh treatment, but will survive long periods in mineral oil, in water, or by frequent transfer; however, most will not survive drying in silica gel or freeze-drying. Entomophthorales do well in liquid nitrogen but will also survive in oil or in soil. Nonsporulating basidiomycetes can be successfully stored in liquid nitrogen. Several other collections of such fungi have been successfully kept alive by frequent transfer, under mineral oil, or by refrigeration (von Arx and Schipper, 1978) and they do well in water or in soil (Smith and Onions, 1983a).

Rusts do not normally grow in culture but living collections can be maintained in good condition in liquid nitrogen (Holden and Smith, 1992). Smuts grow poorly and survive best in liquid nitrogen, although it is possible to keep them by other means, e.g., harvesting spores from infections of the host and freeze-drying. There are fungi, such as species of *Fusarium*, that change or deteriorate rapidly when grown and regularly subcultured. These fungi should be kept in a manner that will obviate serial transfer. Silica gel, water, and soil storage are simple techniques to use if freeze-drying or liquid nitrogen is not available.

Continuous Subculture

The most successful preservation results are obtained when the culture is in an optimum condition of growth at the time of treatment. The major factors affecting growth are medium, temperature, light, aeration, pH, and

TABLE 2
Preservation Techniques for a Variety of Fungi Used at International Mycological Institute

Fungi	Transfer on agar			Water	Deep freeze	Soil	Silica gel	Freeze-drying	Liquid nitrogen
	Room temp	Refrigerator	Mineral oil						
Mastigomycotina (excluding Oomycetes)	Fair	Fair[b]	Fair	Poor	Poor	Fair	Fail	Fail	Fair
Oomycetes	Poor	Poor	Fair/good	Good	Poor	Fair	Fail	Poor[c]	Good
Zygomycotina (excluding Entomophthorales)	Good	Fair[b]	Fair	Good	Fair	Good	Good	Very good	Very good
Entomophthorales	Poor	Poor	Good	Fair	Poor	Fair	Fail	Fail	Good
Ascomycotina (excluding Laboulbeniales)	Good	Good	Fair	Good	Good	Good	Very good	Very good	Very good
Laboulbeniales[a]	Poor	Poor	Poor	Poor	Poor	Poor	Fail	Poor	Good
Basidiomycotina Mycelial in culture	Fair	Good	Good	Good	Good	Good	Fail	Fail	Good
Those that sporulate in culture	Fair	Fair	Good	Good	Good	Good	Good	Good	Very good
Uredinales (rusts)[a]	Fail	Fail	Fail	Fail	Poor	Poor	Fail	Poor	Good
Ustilaginales (smuts)[a]	Fair	Fair	Fair	Fail	Poor	Poor	Fail	Fair	Good
Deuteromycotina	Fair	Good	Poor/good (variation)	Good	Good	Good	Very good	Very good	Very good

[a] Experience with these groups at IMI is limited.
[b] Some fail.
[c] Some species of *Pythium* and *Phytophthora* survive.

TABLE 3
A Comparison of Preservation Methods

Method of preservation	Cost		Longevity	Genetic stability
	Material	Labor		
Serial transfer on agar				
Storage at room temperature	Low	High	1–6 months	Variable
Storage in the refrigerator	Medium[a]	High	6–12 months	Variable
Storage under oil	Low	Low/medium	1–40 years	Poor
Storage in water	Low	Low/medium	2–5 years	Moderate
Storage in the deep freeze	Medium[a]	Low/medium	4–5 years	Moderate
Drying				
In soil	Low	Medium	5–25 years	Moderate to low
Silica gel	Low	Medium	5–19 years	Good
Freeze-drying	High	Initially[b]	4–40 years	Good / Medium
Freezing				
Liquid nitrogen	High	Low	Infinite	Good; 23 years to date at IMI

[a] Refrigerator or deep freeze costs included.
[b] Initial processing is costly depending on the method; maintenance is negligible.

water activity. The importance of starting out with a pure, healthy culture in good condition is to ensure that the correct culture is preserved with all its morphological and physiological characteristics intact. Deteriorated cells may die during the preservation method, so the culture should be in optimum condition when processed. Ideally, cultures should be from freshly isolated material, sporulating and growing on freshly prepared media. A knowledge of the original habitat can give an indication of suitable growth conditions. The type of inoculum can affect the quality of the culture. It is thought that a mass spore transfer is by far the best technique (Onions, 1971).

In contrast, an unhealthy culture may be one that has been kept a long time on stale agar before preservation. This may have caused the culture to change both its morphology and physiology. The culture can form sectors of morphological variants on agar and may also lose the ability to sporulate, or in the case of pathogens the ability to infect a host. This is often due to the culture adapting to growth on synthetic media.

A deteriorating culture can sometimes be revived. To encourage sporulation the culture can be transferred to a medium with limited nutrients, such as potato/carrot agar, and illuminated with near-ultraviolet light (wavelength 200–300 nm, black light). Single-spore cultures (Gordon, 1952) may be used with some unstable organisms; individual spores are transferred to fresh medium and the most "typical" strain is selected from the resulting cultures. This technique is also useful when separating an isolate from a more rapidly growing contaminant.

GROWTH CONDITIONS

The growth requirements for fungi may vary from strain to strain, although cultures of the same species and genera usually grow well on similar media. A medium that induces good sporulation and minimal mycelium formation is most desirable (Smith and Onions, 1983a). The source of isolates can give an indication of suitable growth conditions; thus isolates from environments of low water activity can be expected to grow well on high-sugar media, species from leaves may sporulate best in light, those from salt water may require salt, and those from deserts and the tropics may require high growth temperatures. Cultures are usually best grown on agar slopes in test tubes or culture bottles and, for the short term, *in* petri dishes. A list of recommended media and growth temperatures for common species is given in Smith and Onions (1983a). The majority of fungi can be maintained on a relatively small range of media. However, some fungi deteriorate when kept on the same medium for prolonged periods, so media should be alternated from time to time. If sporulation begins to decrease, transferring the fungus onto a medium with minimal nutrients, such as potato/carrot agar may induce sporulation (Smith and Onions, 1983a).

The majority of filamentous fungi are mesophilic, growing at temperatures within the range of 10–35°C, with optimum temperatures between 15 and 30°C (Smith and Onions, 1983a). Some species (e.g., *Aspergillus fumigatus* and *Talaromyces avellaneus*) are thermotolerant and will grow at higher temperatures. A small number (e.g., *Chaetomium thermophilum, Penicillium dupontii,* and *Thermoascus aurantiacus*) are thermophilic and will grow and sporulate at 45°C or higher, but fail to grow below 20°C. A few fungi (e.g., *Hypocrea psychrophila*) are psychrophilic and are unable to grow above 20°C, whereas many others (e.g., a wide range of *Fusarium* and *Penicillium* species) are psychrotolerant and are able to grow both at freezing point and at room temperature.

Many species grow well in the dark, but others require light and some sporulate better under near-ultraviolet light (Leach, 1962). These latter fungi must be grown in plastic petri dishes or plastic universal bottles for 3–4 days before irradiation. Glass is not suitable because it may be opaque to ultraviolet light. As previously stated, rich growth media should be avoided because they may give rise to excessive growth of mycelium; nutritionally weak media are more suitable for inducing sporulation. Most leaf- and stem-inhabiting fungi (e.g., *Colletotrichum coccodes, Fusarium oxysporum,* and *Glomerella cingulata*) are light sensitive and require light stimulation for sporulation.

Nearly all fungi are aerobic and, when grown in tubes and bottles, obtain sufficient oxygen through cotton wool plugs or loose bottle caps. Care should be taken to see that bottle caps are not screwed down tightly during the growth of cultures. A few aquatic hyphomycetes require additional aeration, by bubbling air through liquid culture media.

Filamentous fungi are variable in their pH tolerance. Most common fungi grow well between pH 3 and 7, although some can grow at pH 2 and below (e.g., *Moniliella acetoabutans, Aspergillus niger,* and *Penicillium funiculosum*).

All organisms need water for growth, but the amount required varies widely. Although the majority of filamentous fungi require high levels of available water, the limiting level of a_w for growth is generally considered to be about 65% relative humidity (Onions *et al.,* 1981). Some xerophilic species are able to grow at low water activity (e.g., *Eurotium* species and *Xeromyces* bisporus). Others that occur on preserves or salt fish will grow well only on media containing high concentrations of sugar or salt (e.g., *Eurotium rubrum, Penicillium zonatum,* and *Wallemia sebi*).

Cultures growing on agar are subject to infestation by mites, commonly of the genera *Tyroglyphs* and *Tarsonemus.* They may damage and/or destroy cultures. They can also introduce other microorganisms carried on their bodies, causing severe cross-contamination that may overgrow and prevent the reisolation of the strain being preserved. Mites can be detected by scrutiny of cultures at twice weekly intervals. The mites can be seen as white objects, just detectable with the naked eye. Ragged colony margins or growth of contaminant fungi or bacteria in the form of trails across the agar may denote their presence. If mites are detected, contaminated cultures should be destroyed by autoclaving, 121°C for 15 min.

To prevent mite invasion, all work surfaces must be kept clean and cultures protected from aerial infestation by storage in cabinets or incubators. To clean work surfaces, wash with 70% ethanol or with the nonfungicidal acaricide, 0.2% (v/v) Actelic (West Care Group, Aldershot, United Kingdom). A cigarette paper adhered to the necks of universal bottles or tubes provides an effective barrier against mites. The cigarette papers are sterilized by dry heat at 180°C for 2 hours and attached by rotating the flamed bottle neck in copper sulfate/gelatin glue (20 g gelatin, 2 g copper sulfate in 100 ml distilled water) and placing the paper on the neck. The excess paper can then be flamed away (Snyder and Hansen, 1946).

If mite-infested cultures cannot be reisolated, they can be stored at −18°C for 1–3 days before being subcultured. This procedure will kill both eggs and adult mites. Fungi that do not survive short-term cold storage may be covered with a layer of mineral oil and subcultured after 24 hours, although this procedure does not kill the mite eggs. However, it must be remembered that the mite will have carried in contaminants and the culture will require decontamination.

SERIAL TRANSFER

The simplest method of maintenance of living fungi is by serial transfer from old media to fresh, on solid or liquid media, and storage in the most suitable conditions for the individual isolate. Many fungi can be maintained in this way for years by growth on suitable media. Successful maintenance is dependent on transfer from the growing margins of the culture, taking care to ensure that contaminants or genetic variants do not replace the original strain. Certain groups of fungi have specific growth requirements. Some dermatophytes survive best on hair (Al-Doory, 1968), some water molds are best stored in water with the addition of plant material (Goldie-Smith, 1956), and other more sensitive water molds may require aeration (Clark and Dick, 1974; Webster and Davey, 1976).

The main disadvantages of this method are the danger of variation, loss of pathogenicity or other physiological or morphological characteristics, and the possibility of contamination by airborne spores or mite-carried infections. Serial transfer requires constant specialist supervision to ensure that the fungus is not replaced by a contaminant or subcultured from a deteriorated sector.

The time period between transfers varies from fungus to fungus, for some every 2–4 weeks, the majority every 2–4 months, though others may survive for 12 months without transfer. The use of cold storage can slow the rate of metabolism and thus increase the intervals between subcultures. Storage at 4–7°C in a refrigerator or cold room can extend the transfer interval to 4–6 months from the average period of 2–3 months. Some fungi are sensitive to storage at these temperatures, particularly thermophiles, and many representatives of the Oomycota. Overpacking of refrigerators can cause a buildup of condensation that may promote cross-contamination. Storage in a deep freeze (−17 to −24°C) will allow most fungi to survive 4–5 years between transfers, though freezing damage may occur with some. Mites do not invade cultures at these lower temperatures.

STORAGE UNDER MINERAL OIL

Covering culture slants with mineral oil prevents dehydration and slows down the metabolic activity through reduced oxygen tension. This method was first extensively used by Buell and Weston (1947), and subsequent reports have indicated its wide application and success (Onions, 1977; Smith, 1988).

Mature healthy cultures in 30 ml universal bottles are covered by 10 mm of mineral oil (liquid paraffin or medicinal paraffin specific gravity 0.830–0.890) that has been sterilized by autoclaving twice, 24 hours apart, at 121°C for 15 min. If the depth of oil is greater the fungus may not receive sufficient oxygen and may die. If the depth is less, exposed mycelium or agar on the side of the container may allow moisture to evaporate and the culture to dry out. It is important when subculturing from stored cultures that as much oil is drained from inocula as possible. More than one subculture may be necessary after retrieval because the growth rate can very often remain reduced due to adhering oil. The fungal mycelium can normally recover when it is reisolated from the agar plate onto fresh media. There is an added risk of oil, containing fungus, contaminating the worker when sterilizing inoculation needles in the bunsen flame.

A wide range of fungi survive this method. Saprolegniaceae and other water molds survived 12–30 months (Reischer, 1949), and species of *Aspergillus* and *Penicillium* have remained viable for 32 years (Smith and Onions, 1983a). However, many cultures deteriorate under a layer of mineral oil (Smith and Onions, 1988b). Some species of *Cercospora, Arthrobotrys, Colletotrichum, Conidiobolus, Corticium,* and *Nodulisporium* should be transferred every 2 years.

The disadvantages of oil storage are contamination by airborne spores, retarded growth on retrieval, and the presence of oil in microscope preparations, which impedes or spoils observation. Additionally, adhering oil can contaminate the outside of bottles, their caps, and personnel handling them. Continuous growth of the culture under adverse conditions places the fungus under selection pressure. Any one of many variants occurring in the population of generated spores may grow more profusely under the conditions provided.

WATER STORAGE

Species of *Phytophthora* and *Pythium* have been stored for periods of 2–3 years before any loss of viability was noted (Onions and Smith, 1984). These cultures showed some deterioration in pathogenicity, but the majority were able to infect their host. Viability deteriorated rapidly after 2 years of storage and 42% (21/50) of the isolates were dead at 5 years.

The method of water storage was orginally described by Castellani (1939, 1967), who stored fungi pathogenic to humans. Figueiredo and Pimentel (1975) reported 10 years of successful storage of plant pathogens without loss in pathogenicity. Boeswinkel (1976) stored 650 plant pathogens from a variety of fungal groups. They all remained viable and pathogenic for 7 years. Clark and Dick (1974) reported successful results with Oomycota; Ellis (1979), with Entomophthorales, pyrenomycetes, hymenomycetes, gasteromycetes, and hyphomycetes; and Marx and Daniel (1976), with ectomycorrhizal fungi.

Growth may sometimes occur during storage in water. This will be reduced if the spores or hyphae are removed from the surface of agar media and no medium is transferred. Agar blocks, 6 mm^3, are cut from the growing edge of a fungus colony and are placed in sterile distilled water in McCartney bottles, with the lids tightly screwed down, and stored at room temperature.

Dehydration

Most fungal spores have a lower water content than do vegetative hyphae and are able to withstand desiccation. Dehydration reduces the environment in which chemical reactions occur and thus reduces metabolism. Only when water is restored will the fungus revive and grow. Several preservation techniques rely on drying: air drying, adding fungi to soil or silica gel, absorbing suspensions into filter paper, drying under vacuum from the liquid state (L-drying), and freeze-drying. Many *Aspergillus* and *Penicillium* species survive air drying (Smith and Onions, 1983a) and *Allomyces arbuscula* has been reported to survive this method for up to 17 years (Goldie-Smith, 1956). The spores of vesicular–arbuscular endomycorrhizas (e.g., *Glomus, Acaulopsora,* and *Gigaspora*) have been preserved by drying at 22°C under vacuum (Tommerup and Kidby, 1979).

SILICA GEL STORAGE

Storage in silica gel has proved to be very successful at IMI, where sporulating fungi have been stored for 7–19 years and have recovered well, remaining morphologically and genetically stable (Smith, 1988). The technique is to inoculate cold silica gel with a suspension of fungal propagules and then dehydrate them to enable their storage without growth or metabolism. Glass bottles are one-quarter filled with nonindicator silica gel and sterilized in an oven (180°C for 3 hours). The bottles are then placed in a tray of water, which is frozen in a deep freeze ($\sim -20°C$) and stored overnight. The tray is removed from the deep freeze 20 to 30 min before required. Spore suspensions are made in cooled sterile 5% (w/v) skimmed milk (5°C); this is then added to the silica gel crystals so that it percolates three-quarters of the way down through them. The bottles are freed from the ice and shaken to agitate the crystals to ensure that they are coated with spores. After 10–14 days of incubation at 25°C, when the crystals readily separate, the bottle caps are screwed down tightly and are stored at 4°C in airtight containers. Indicator silica gel is added to the outer container to absorb moisture. A wide range of sporulating fungi survive this method (Perkins, 1962; Ogata, 1962; Gentles and Scott, 1979; Smith and Onions, 1983b). Thin-walled spores and spores with appendages and mycelium, e.g., zoospores and oogonia of the oomycetes *Phytophthora* and *Pythium,* the appendaged spores of *Bartalinia,* and the long filiform spores of some *Cercospora* species, tend not to survive this technique (Smith and Onions, 1983a).

SOIL STORAGE

Soil storage involves inoculation of 1 ml of spore suspension into soil that has been autoclaved twice, 24 hours apart, at 121°C for 15 min, and incubating at room temperature for 5–10 days. This initial growth period allows the fungus to use the available moisture and gradually to become dormant. The bottles are then stored in a refrigerator (4–7°C). This method of storage has proved very successful with *Fusarium* species (Booth, 1971).

Atkinson (1953, 1954) found that species of *Rhizopus, Alternaria, Aspergillus, Circinella,* and *Penicillium* survived 5 years of storage in soil without any morphological and physiological changes. Shearer *et al.* (1974), working with *Septoria,* and Reinecke and Fokkema (1979), working with *Pseudocercosporella herpotrichoides,* have reported stable and successful results.

FILTER PAPER DISC

Yeasts have been dried on filter paper and stored above the desiccant silica gel (Kirsop and Kurtzman, 1988). Squares or discs of Whatman No. 4

filter paper 10 mm across are placed in an aluminum foil packet and sterilized by autoclaving. The discs or squares are inoculated by immersing into drops of yeast suspension prepared in 5% (w/v) skimmed milk. The inoculated filter paper is dried in a desiccator in the foil packets and stored in an airtight container at 4°C. Cultures are revived by removing a piece of filter paper and placing it on an agar plate or in broth medium and incubating at a suitable temperature. In general cultures should be replaced at intervals of 2–3 years, thus allowing the successful preservation of haploid and diploid strains of *Saccharomyces cerevisiae* with a wide variety of genetic markers (Kirsop and Doyle, 1991). However, poor revivals have been obtained with *Yarrowia lipolytica* (Kirsop and Doyle, 1991).

L-DRYING

L-Drying is a method in which the organism is dried under vacuum without freezing (Tommerup and Kidby, 1979). A method described by Malik (1991) has been used successfully to store representatives of groups that fail the simple techniques, e.g., under liquid paraffin, distilled water, drying on silica gel, and drying on glass beads. Generally suspensions are made in 20% (w/v) skimmed milk and are supported on a carrier such as a fiber filter or neutral activated charcoal and a protective agent such as inositol or glutamate (5%, w/v). The mixture is placed in 0.5-ml neutral glass ampules and attached to a vacuum system with a cold trap to remove the evaporating water. Water is removed holding the mixture at room temperature. A secondary stage may follow wherein drying is carried out over a desiccant. The ampules are then sealed to retain low pressure. Organisms are generally recovered by opening the ampule and adding liquid growth medium. This method can be used successfully for some yeasts and many filamentous fungus spores (Tommerup and Kidby, 1979).

FREEZE-DRYING (LYOPHILIZATION)

Preservation by drying under reduced pressure in the frozen state by the sublimation of ice was first extensively used with fungal cultures by Raper and Alexander (1945). The methods and machinery have been improved over the years to produce a reliable and successful preservation technique for sporulating microfungi. Stability and long storage periods have been shown to be the main advantages of freeze-drying (von Arx and Schipper, 1978; Jong *et al.,* 1984).

The basic requirements for a freeze-drying system are a means of freezing the suspension, of generating and maintaining a vacuum, and of absorbing the water vapor evolved. To freeze a culture, ampules containing a spore suspension can be attached to a vacuum pump via a desiccant trap and immersed in a "freezing bath." The ampules can then be evacuated. Drying is more rapid once the ampules have been removed from the freezing mixture, although care must be taken not to allow their contents to thaw. A vacuum gauge is useful to monitor the process, giving an indication of the amount of water evaporating in the system and therefore the degree of dryness of the product. The equipment used is limiting and the method relies entirely on it. Reports stating whether freeze-drying is successful must be related to the procedure by which drying was achieved. The protectant used, rate of cooling, final temperature, rate of heat input during drying, residual moisture, and storage conditions all affect the viability and stability of fungi (Smith, 1986).

The suspending medium should be convenient to use and give protection during the process—i.e., protect the spores from freezing damage and storage problems such as oxidation. Media used are skimmed milk, serum, peptone, and various sugars or mixtures of them (Table 4). Some sugars have associated problems due to their behavior during freeze-drying. Glucose and nutrient broth solutions, for example, can glaze and the skin formed can prevent proper drying. Bubbling can also occur prior to freezing or later in the process if the solutions are allowed to thaw. This gives reduced viabilities after relatively short storage periods because of the damage incurred during drying and the high residual water content increasing the rate of deterioration. Many protectants are sensitive to heat and denature easily. Skimmed milk, sugars and peptone are sterilized by autoclaving at 114°C for 10 min. However, serum and other heat-sensitive substances are usually sterilized by filtration.

The rate of freezing is a very important factor that must be considered in order to get optimum recovery. Slow rates of freezing have been successful with fungi, with $-1°C/min$ the rate normally quoted as best (Hwang, 1966; Heckly, 1978). The technique of centrifugal freeze-drying relies on evaporative freezing under reduced pressure, which gives a favorable rate of cooling and can be used successfully for the storage of many sporulating fungi (Smith, 1983a). However, the mechanism may be damaging in itself. More rapid cooling by immersion in freezing mixtures may give poorer viabilities, but a technique that allows the variation of cooling rate to suit the organism being freeze-dried, such as in a shelf freeze-drier, is preferable. The use of a programmable cooler to control the cooling of the organism in suspension and transfer to the precooled shelves of a freeze-drier allows the development of suitable protocols for individual strains.

Overdrying will kill or in other cases cause mutation by damaging DNA (Ashwood-Smith and Grant, 1976). The residual moisture content must not be allowed to fall below 1% (Smith, 1988).

Storage of the freeze-dried material should exclude the presence of both oxygen and water vapor because they can cause rapid deterioration (Rey, 1977). This can be achieved by filling the ampules or vials with a dry inert gas, such as nitrogen or argon, allowing the gas to flow into the freeze-drier through the air admittance valve before sealing the ampules. Storage of ampules with an internally reduced pressure can also reduce deterioration if the ampule has an airtight seal. Storing the ampules at low temperature (4°C) is recommended for greater longevities (Heckly, 1978; Jong et al., 1984). However, ampules can be stored at room temperature for over 25 years without significant loss in viability (Smith, 1983a).

Rehydration of the fungi with sterile distilled water should be carried out slowly, allowing time (~30 min) for absorption of moisture before plating onto a suitable medium. If possible, rehydration should be carried out in a controlled environment for especially sensitive strains by placing the organism in an environment of high humidity or placing it in an isotonic medium. It is advisable to rehydrate and check viabilities on a regular basis, immediately after preservation and at 10-year intervals or more frequently if viability is low. Cultures should not be placed on media that put the organism under stress, i.e., media with low water activity, high salt content, or inhibitors.

Studies at IMI have shown that many strains failed to survive the freeze-drying; 45% died during the freezing stages. However, improving cooling protocols alone did not result in a significant improvement in survival. Many strains still fail to survive the dehydration stages under vacuum. A critical examination of the centrifugal freeze-drying technique used at IMI revealed that 1.8% (134 of 7365 strains) died in storage over a period of 25 years (Smith, 1986). To achieve the best recoveries, a moisture content of 5% by dry weight should be reached before the temperature of the sample is allowed to rise above $-15°C$. Allowing temperature to rise above this while water content remained high resulted in melting of the cytoplasm and in the fungus drying while in the liquid state, and not while frozen, which gave reduced viabilities. The final residual moisture content was also crucial. Residual water contents of over 10% resulted initially in high viabilities with *Aspergillus niger* and *Penicillium ochrochloron.* However, following 1 year of storage only 55% of *Aspergillus* propagules and <20% of *Penicillium* remained viable (Smith, 1986). It is necessary to determine protocols for each strain to get optimum stability and viability. The suspending medium, the cooling rate, the temperature where drying begins (ensuring that it remains below the freezing point of the fungus during initial drying), and final residual moisture must be optimized to perfect the storage conditions. Cooling at $-1°C/min$ in a 10% skimmed milk and 5% inositol mixture, holding the temperature below $-20°C$ for the first 3 hours of drying, enabled the successful preservation of the hyphae of *Armillaria mellea.* The residual moisture being ~2% (Smith, 1986). Those organisms

TABLE 4

Cryoprotectants Used in Freezing and Freeze-Drying: Advantages and Disadvantages

Cryoprotectant	Method	Advantages	Disadvantages	Organisms surviving	References
Agar powder	Liquid nitrogen storage	Good dry support for leaf infections of obligate pathogens	Low viability	*Plasmopara viticola, Pseudoperonospora humuli*	Tetsuka and Katsuya (1983)
Beef serum	Freeze-drying	Simple to prepare	Not a defined medium; may contain toxic virus; large ascospores fail	Many species of *Aspergillus* and *Penicillium*; representatives of Mucorales, *Rhizopus* and *Phycomyces*; and the deuteromycetes	Raper and Alexander (1945)
Dimethyl sulfoxide (DMSO 5% v/v)	Liquid nitrogen storage	Gives high viability; penetrates cell quickly	Toxic to microorganisms and to humans; will not freeze-dry	Ascomycota (44 strains), sterile mycelium. Entomophthorales (19 strains), deuteromycetes(33 strains), Ascomycota (62 strains). deuteromycetes (239 strains), *Coprinus cinereus, Basidiobolus* spp.. *Aspergillus* (6 spp.. 8 strains)	Hwang (1966) Butterfield *et al.* (1974) Xie Yi-mei and Cong Shao-hai (1987) Long *et al.* (1978) Yarlett *et al.* (1986) Li Zhong-ging and Yuan Wei (1984)
DMSO (10% v/v)	Liquid nitrogen storage	Gives high viability; penetrates cell quickly	Toxic to microorganisms and to humans; will not freeze-dry	*Sclerospora phillippinensis, Sclerospora sacchari, Neocallimastix patriciarum;* yeasts (33 strains), including *Candida lipolytica, Geotrichum candidium, Saccharomyces cerevisiae, Schizosaccharomyces octosporus*	Dahmen *et al.* (1983)
DMSO (15% v/v)	Liquid nitrogen storage	Gives high viability; penetrates cell quickly	Toxic to microorganisms and to humans; will not freeze-dry	*Bremia lactucae, Plasmopara viticola, Sclerospora sorghi, Peronospora tabacina*	Gale *et al.* (1975); Bromfield and Schmitt (1967)
DMSO (10% v/v) and glucose (8% v/v)	Liquid nitrogen storage	Lower toxicity and better viability than DMSO alone	Some toxicity	Deuteromycetes (9 strains), *Phytophthora palmivora, Pythium sylvaticum, Pseudophaeolus boudonii*	Smith (1983b)
Glucose broth (7.5% w/v glucose) and Oxoid nutrient broth No. 2 (2.5% w/v)	Freeze-drying	Good universal medium for bacteria	Can form skin, which impedes drying	Bacteria; fungi survive less well and die in storage	Lapage and Redway (1974)

Additive	Method	Advantages	Disadvantages	Organisms	Reference
Glycerol 10% v/v	Liquid nitrogen storage	Widely applicable	Low toxicity	*Achyla* spp., *Allomyces* spp., *Aphanomyces* sp., Ascomycota (62 strains), *Basidiobolus* sp., deuteromycetes (239 strains). *Coprinus cinereus*, deuteromycetes (33 strains), Entomophthorales (19 strains), *Hyphochytrium catenoides*, (sterile mycelium); >2000 strains of 635 genera of fungi, including *Pythium* spp., *Phytophthora* spp., *Thraustotheca clavata*	Hwang (1966); Smith (1982) Butterfield *et al.* (1974) Hwang (1966) Smith (1982)
Glycerol (10% v/v) and skimmed milk (8.5% w/v)	Liquid nitrogen storage	Improved viability for some fungi over glycerol alone	Low toxicity	*Plasmopara viticola*, *Phytophthora* spp. (3 strains), *Pythium ultimum*,	Dahmen *et al.* (1983)
Inositol (7.5% w/v)	Freeze-drying	A defined medium	Some species fail	yeasts	von Arx and Schipper (1978)
Poly(vinyl alcohol)	Liquid nitrogen storage	Reduces ice crystal size	Toxic to some cells	*Plasmopara viticola*, *Pseudoperonospora humuli*	Tetsuka and Katsuya (1983)
Polyvinylpyrrolidone	Liquid nitrogen storage	Minimizes contact with ice	Does not penetrate the cell; does not protect representatives of the Oomycota; difficult to sterilize; viscous and heat sensitive	*Aspergillus carbonarius*, *Cercospora xanthosomatis*, *Cunninghamella elegans*, *Drechslera* sp., *Emericellopsis terricola*, *Penicilliopsis clavariiformis*, *Phycomyces blakesleeanus*, *Wallemia sebi*	Smith (1983b)
Skimmed milk (10% w/w)	Freeze-drying	Simple to prepare	Not a defined medium; intricate spores or spores with delicate walls fail	Fungal spores, including *Mucor*, *Mortierella*, *Neurospora*, *Phycomyces*, *Rhizopus*, and *Streptomyces*	von Arx and Schipper (1978)
Skimmed milk	Freeze-drying	Simple to prepare	Not defined	Fungi	Alexander *et al.* (1980)
Skimmed milk (10% w/v) and inositol (5% w/v)	Freeze-drying	The addition of inositol acts as a water buffer, preventing over drying	—	>11,000 strains of 3500 species of fungi	Smith (1991)

that have survived at IMI have shown no detectable change in morphological or biochemical characteristics.

Preservation by freeze-drying facilitates the distribution of preserved fungi. Ampules can be taken from storage, packaged according to postal regulations, and dispatched. This rapid response to culture requests is essential for culture collections. It also allows the organism to withstand adverse transport conditions and delays that would normally promote deterioration of growing strains.

Cryopreservation

Lowering the temperature of living cells reduces the rate of metabolism until, when all internal water is frozen, no further biochemical reactions occur and metabolism is suspended (Franks, 1981). Although little metabolic activity takes place below $-70°C$, recrystallization of ice or ice crystal growth can occur at temperatures above $-139°C$ (Morris, 1981), and this can cause damage during storage. The volume occupied by water increases by 10% when water crystallizes and forms ice. This puts the cell under mechanical stress (Grout and Morris, 1987).

The storage of microorganisms at ultralow temperatures ($-196°C$ for liquid nitrogen) is at present regarded as the best method of preservation (Kirsop and Doyle, 1991; Smith, 1988). It is a form of induced dormancy, during which the organism does not undergo any change either phenotypically or genotypically, providing adequate care is taken during freezing and thawing. The method can be applied to both sporulating and nonsporulating cultures. Optimization of the technique for individual strains has enabled the preservation of organisms that have previously failed (Morris *et al.,* 1988).

At IMI over 4000 species belonging to over 700 genera have been successfully frozen in 10% (v/v) glycerol. No morphological or physiological changes have been observed (Smith and Onions, 1983a). The original method involved preparation of a suspension of fungal spores or mycelium in a protective medium, freezing at $-1°C/min$ to $\sim-35°C$, followed by rapid uncontrolled cooling to $-196°C$; recovery was achieved by thawing rapidly in a water bath at 35°C. Variations of the method are numerous. Many different rates of cooling and thawing and the use of different cryoprotectants have been attempted by many workers (Smith, 1983b; Morris *et al.,* 1988). Initial work with fungi was undertaken by Hwang (1960) at the American Type Culture Collection. Further reports of methods used have been given (Calcott, 1978; Onions, 1983; Smith, 1988).

Both sterile and sporulating cultures survive the technique; weak isolates may be more sensitive to the process (Smith and Onions, 1983a). Care must be exercised to prevent mechanical damage during preparation of suspensions and various precautions to avoid this can be taken. For example, fungi on slivers of agar can be put into the ampules or the fungus can be grown on small amounts of agar in the ampule before the cryoprotectant is added.

The choice of cryoprotectant is a matter of experience and varies according to the organism being preserved (Table 4). Glycerol often gives vary satisfactory results, but needs time to penetrate the organisms; some organisms are damaged by this delay. Dimethyl sulfoxide (DMSO) penetrates rapidly and is often more satisfactory (Hwang and Howells, 1968; Hwang *et al.*, 1976). Other protectants have been used, including sugars and large molecular substances such as polyvinylpyrrolidone (PVP) (Ashwood-Smith and Warby, 1971). Mixtures of protectants have proved particularly satisfactory at IMI (Smith, 1983b). In some instances the presence of such an additive is detrimental. Urediniospores of *Puccinia abrupta* var. *partheniicola* survived with 85% viability and remained infective, whereas in the presence of glycerol, DMSO, PVP, or trehalose viability and infectivity decreased (Holden and Smith, 1992).

Slow thawing may cause damage due to recrystallization of ice during warming, therefore rapid warming is recommended. Slow freezing and rapid thawing generally give the highest viabilities (Heckly, 1978).

Storage is at $-196°C$ in the liquid nitrogen or slightly higher in the vapor phase immediately above it. However, temperatures can rise to $-90°C$ under the lid of the storage vessel and to higher temperatures when the vessel is open for retrieval or freezing of organisms. The development of evaporating freezers that can produce temperatures of $-150°C$ or lower may result in their use in the future if they prove reliable and durable.

Choosing Preservation Techniques

The choice of methods will depend on the requirements of the collection and the equipment and facilities available. Table 3 compares different methods of preservation with regard to costs of materials and labor, longevity, and genetic stability. It is recommended that each isolate should be maintained by at least two different methods.

In general, freezing and storage in liquid nitrogen is the best preservation technique available for filamentous fungi (Table 2). The handling techniques, freezing protocol, cryoprotection, and thawing rates can be optimized for a particular fungus to obtain maximum survival. Once the fungus has been successfully frozen and stored in liquid nitrogen, the storage periods appear to be indefinite, because no chemical and very few physical changes can occur at such low temperatures (Grout and Morris, 1987).

However, the problem of ice crystal growth during thawing remains (albeit small at rapid rates of warming).

Freeze-drying is recommended as a second choice, although generally only the sporulating fungi survive. Freeze-dried ampules are more convenient because they can be stored at room temperature and require little maintenance. A Commission of the European Community Biotechnological Action Programme project to improve freeze-drying and cryopreservation confirmed the value of these techniques (Smith *et al.*, 1990a,b). Optimization of the techniques was necessary to ensure stability of properties. Silica gel storage can be recommended for the laboratory with limited resources, because it also prevents genetic drift. Other methods are less valuable because they allow variation, although they can be used to back up other techniques (Table 3).

Evaluating Viability

The viability of fungi is assessed by a variety of different means depending on the structures preserved. Germination of the fungus spore is the simplest method. This can be observed directly on the agar medium on which the spore has been inoculated, or on a microscope slide. Normally, dilutions of harvested spores or resuscitated preserved spores must be prepared because, in concentrated preparations, germination of one spore may be impeded by another in close proximity. Generally a concentration below 1×10^3 ml^{-1} should be used. Germination tests are carried out by placing a drop of spore suspension on a microscope slide, immediately examining under the microscope for spores that have already germinated, and incubating them overnight at 25°C. The humidity is kept high by placing the microscope slide on a bent glass rod in a petri dish containing moist filter paper. The slide is reexamined and the proportion of germinated spores is counted for at least 10 microscope fields at a magnification of 400× to give a total of at least 100 spores counted where possible.

Alternatively a drop of the diluted spore suspension (prepared as above) is streaked across a petri dish of tapwater agar (TWA), malt agar (MA), or potato/carrot agar (PCA) and is examined under the microscope for spores that have germinated prior to incubation. The plates are incubated at 25°C and examined at suitable intervals up to a total ~7 days. The percentage of germinated spores is calculated for at least 10 microscope fields. When the spore concentration in the suspension is low, spores are examined along the streaks on the agar at magnification 400× to give a total of approximately 100 or more spores counted where possible. All spore germination tests must be followed by assessing rate of growth, colony morphology, structures, and pigmentation as compared to the original cultures, because in some cases spore germination may be abortive.

Determining the viability of nonsporulating fungi is a more difficult task. Macerated cultures are predisposed to further damage during preservation and a loss in viability. To avoid this agar, plugs can be used. To test viability, place on growth media; the proportion giving rise to fresh growth can be determined. Another method quite often used is to grow the fungus in agitated liquid medium to produce small, intact colonies, 0.5 to 1.0 mm in diameter, for preservation. The viability of these can be determined by inoculating directly onto agar media and examining immediately and at intervals for growth of the organism (for example, 36, 48, and 72 hours for *Penicillium* species and 2–7 days for some species of *Phytophthora* and some representatives of the Basidiomycota). Obligate plant pathogens can be assessed by their ability to germinate and infect their host whenever possible. Obviously some hosts are not always available. Additionally nonindigenous and other plant pathogens must be contained, and government permits are necessary.

Special Methods and Considerations for Recombinant Fungal Vectors and Hosts

Cloning genes and transferring them between organisms have become common in many laboratories. The objective, until recently, has been to examine existing systems rather than to create new organisms. However, the construction of new organisms for biotechnology is attracting interest. Mutants have been produced in the past, but it has become more effective to introduce specific genes into organisms that have preferable growth characteristics. It is now possible to identify and isolate many genes. Cloning genes from yeast cells has been accomplished and the most commonly used vector for these is the bacterial plasmid (Fincham and Ravetz, 1991). Genetic engineering using filamentous fungi has been discussed by Bennett and Lasure (1985). An example of successful genetic manipulation in the filamentous fungi is the gene for the enzyme endoglucanase I, which has been isolated from *Trichoderma reesei* and cloned and expressed in the yeast *Saccharomyces cerevisiae* (van Ardsell *et al.,* 1987). Techniques that allow continuous growth or put the organism under selective pressure may cause the loss of the plasmid and therefore the desired property.

Yeasts

Reports of the loss of expression of plasmids and the induction of respiratory-deficient mutants in yeasts caused by freeze-drying and poor cryopreservation techniques necessitate careful monitoring of plasmid-bearing organisms (Pearson *et al.*, 1990). A study of plasmid expression in *S. cerevisiae* following freezing and thawing revealed that yeast plasmid DNA was permanently, lost following fast cooling rates (Pearson *et al.*, 1990). It is necessary to avoid the extreme stresses of freezing and thawing protocols to prevent DNA loss. The fast cooling rate of many evaporative cooling procedures in freeze-drying should be avoided to retain stability. Generally, controlled slow linear cooling prior to evacuation will give better results (Pearson *et al.*, 1990). However, it is important that extensive shrinkage of the cell be avoided. This will occur during slow cooling when ice nucleates extracellularly, resulting in an increase in concentration of the suspending medium. As a result, water will be lost, reducing the cell volume. The addition of cryoprotectants will minimize such events but will not prevent them (Morris *et al.*, 1988). Because the cell and nuclear membrane have only limited abilities to decrease in surface area and increase in thickness, invaginations are produced. This will allow the cell to reduce in size. However, if the invaginations result in the loss of membrane material in the form of vesicles, or otherwise, to the cytoplasm or outside the cell on thawing, expansion caused by influx of water may result in cell rupture. Cells may avoid rupture by losing cytoplasmic and nuclear material, with the possible consequence of loss of plasmid expression. Carefully designed protocols of freezing and freeze-drying are generally required to retain plasmid expression within the host. This will necessitate the investigation of the effect of freeze–thaw cycles on the properties of the organism and determination of the optimum procedure for cryopreservation. Following such studies with observations of the response to different warming procedures will help determine optimum freeze-drying protocols. Such studies have been carried out at IMI for the development of methods of preservation for fungi.

Filamentous Fungi

Filamentous fungi also require designed protocols to ensure long-term stability of plasmid expression. Plasmid expression is more often lost following freeze-drying than following cryopreservation to low temperatures; carefully controlled freeze-drying and cryopreservation techniques can optimize plasmid retention. Freeze–thaw cycles that place the cells under osmotic stress or internal freezing may result in the loss of plasmid. Cryogenic light microscopy reveals that although high viability of fungi can be obtained following both fast and slow cooling, the cells may suffer shrinkage or freezing damage (Morris *et al.*, 1988). Metabolism is often disrupted following fast cooling and thawing protocols, and in many cases results in permanent loss of properties. This may be due to the loss of plasmid expression in some cases. The use of the cryogenic light microscope allows the observation of the cell's response to freezing and thawing and thus allows the selection of the least stressful process for preservation. Such studies reduce the time and resources spent trying to determine such conditions on a trial and error basis. Although viability studies and checks on the retention of characteristics must still be carried out, this will be kept to a minimum by testing only those protocols that place the cells under minimum stress.

Culture Repositories

Microbial Resource Collection Information Sources

European Culture Collection Organization (ECCO)
ECCO Secretary
VTT Collection of Industrial Microorganisms
VTT Biotechnical Laboratory
Tietotie 2
SF-02150 ESP00
Finland

Japanese Federation for Culture Collections (JFCC)
Japanese Collection of Microorganisms
The Institute of Physical and Chemical Research (RIKEN)
Hirosawa 2-1, Wako-shi
Saitama 351-01
Japan

Microbial Information Network Europe (MINE)
The Coordinator
Centraalbureau voor Schimmelcultures
Oosterstraat 1
PO Box 273
3740 AG Baarn
The Netherlands

Microbial Resource Centres (MIRCEN)
The MIRCEN Secretariat
Division of Scientific Research and Higher Education
United Nations Educational Scientific and Cultural Organization (UNESCO)
7 Place de Fontenoy
75700 Paris
France

Microbial Strain Data Network (MSDN)
The Secretariat
307 Huntingdon Road
Cambridge CB3 01X
United Kingdom

United Kingdom Federation for Culture Collections (UKFCC)
F. G. Priest (Secretary)
Department of Biological Sciences
Heriot-Watt University
Edinburgh EII14 4AS
United Kingdom

United States Federation for Culture Collections (USFCC)
M. Jackson (Secretary-Treasurer)
Abbot Laboratories
Dept 47P
Bldg. AR9A
Abbott Park, Illinois 60064

World Data Center for Microorganisms (WDCM)
Life Science Research Information Section
RIKEN
2-1 Hirosawa, Wako
Saitama 351-01
Japan

World Federation for Culture Collections (WFCC)
WFCC Secretariat
Deutsche Sammlung von Mikroorganismen und Zelkulturen GmbH (DSM)
Masoheroder Weg 1B
3300 Braunschweig
Germany

The Ten Largest Collections Listed at the World Data Center and Their Holdings (Sugawara, 1993)

Agricultural Research Service Culture Collection (NRRL)
USDA
Peroria, Illinois, 61604
    WDC Number: 97.
    Number of Strains: 78,010.
    Organisms held: Algae, bacteria, fungi, yeasts, and actinomycetes.

American Type Culture Collection (ATCC)
Rockville, Maryland 20852
    WDC Number: 1.
    Number of Strains: 53,615.
    Organisms held: Algae, bacteria, fungi, yeasts, protozoa, cell lines, hybridomas,
        viruses, vectors, plasmids, and phage.

Centraalbureau voor Schimmelcultures (CBS)
Oosterstraat 1
P.O. Box 273
3740 AG Baarn
The Netherlands

WDC Number: 133.
Number of Strains: 41,300.
Organisms held: Fungi, yeasts, lichens, and plasmids.

Culture Collection University of Goteborg (CCUG)
Department of Clinical Bacteriology
Guldhedsg 10
S-413 46, Goteborg
Sweden
WDC Number: 32.
Number of Strains: 28,100.
Organisms held: Bacteria, fungi, and yeasts.

International Mycological Institute (IMI)
Bakeham Lane
Egham, Surrey TW20 9TY
United Kingdom
WDC Number: 214.
Number of Strains: 19,500.
Fungi, bacteria, and yeasts.

Mycoteque de l'Universite Catholique de Louvain (MUCL)
Place Croix du Sud 3
Bte 8
B1348, Louvain-Ln-Neuve
Belgium
WDC Number: 308.
Number of Strains: 15,000.
Organisms held: Fungi and yeasts.

Institute for Fermentation Osaka (IFO)
17-85 Jugo-Honchmachi 2 chome
Yodogawaku Osaka 532
Japan
WDC Number: 191.
Number of Strains: 13,443.
Organisms held: Bacteria, fungi, yeasts, cell lines, and viruses.

Canadian Collection of Fungus Cultures (CCFC)
Canada
WDC Number: 150.
Number of Strains: 10,000.
Organisms held: Fungi.

Fermentation Research Institute (FRI)
Japan
WDC Number: 586.
Number of Strains: 9,800.
Organisms held: Bacteria and fungi.

The Upjohn Culture Collection [UC(UPJOHN)]
USA
  WDC Number: 168.
  Number of Strains: 9,390.
  Organisms held: Algae, bacteria, fungi, yeasts, protozoa, plasmids, hybridoma, and
    actinomycetes.

MIRCEN Network in Environmental, Applied
Microbiological, and Biotechnological Research

BIOTECHNOLOGY

Ain Shamus University
Faculty of Agriculture
Shobra-Khaima, Cairo
Arab Republic of Egypt

Applied Research Division
Central American Research Institute for Industry (ICAITI)
Ave. La Reforma 4-47 Zone 10
Apdo Postal 1552
Guatemala

Department of Bacteriology
Karolinska Institutet, Fack
S-10401 Stockholm
Sweden

Institute for Biotechnological Studies
Research and Development Centre
University of Kent
Canterbury, Kent
CT27TD United Kingdom

International Mycological Institute
Bakeham Lane
Egham, Surrey, TW20 9TY, United Kingdom

FERMENTATION, FOOD, AND
WASTE RECYCLING

Thailand Institute of Scientific and Technological Research
196 Phahonyothin Road
Bangkhen, Bangkok 9
Thailand

FERMENTATION TECHNOLOGY

ICME
University of Osaka
Suitasha 563, Osaka
Japan

FERMENTATION TECHNOLOGY
VETERINARY MICROBIOLOGY

University of Waterloo
Waterloo, Ontario, N2LK 3G1 and University of Guelph
Guelph, Ontario NIG 2W1
Canada

MARINE BIOTECHNOLOGY

Department of Microbiology
University of Maryland
College Park Campus 207742
Maryland

RHIZOBIUM

Department of Soil Sciences Botany
University of Nairobi
P.O. Box 30197
Nairobi, Kenya

IPAGRO
Postal 776
90000 Porto Alegre
Rio Grande de Sul
Brazil

NifTAL Project
College of Tropical Agriculture and Human Resources
University of Hawaii
P.O. Box "O"
Paia, Hawaii 96779

Centre National de Recherches Agronomiques d'Institute Sénéglais de Recherches
Agricoles
B.P. 51 Bambey
Senegal

Cell Culture and Nitrogen-Fixation Laboratory
Room 116
Building 011-A Barc West
Beltsville, Maryland 20705

WORLD DATA CENTRE

RIKEN
Saitama
Japan

Acknowledgments

We thank the CAB International Research and Development Fund and the CEC Biotechnological Action Programme for support of part of the work described here.

References

Al-Doory, Y. (1968). *Mycologia* **60,** 720–723.

Alexander, M. J., Daggett, P.-M., Gherna, R., Jong, S. C., Simione, F., and Hatt, H. (1980). *In* "American Type Culture Collection Methods." ATCC, Rockville, MD.

Ashwood-Smith, M. J., and Grant, E. (1976). *Cryobiology* **13,** 206–213.

Ashwood-Smith, M. J., and Warby, C. (1971). *Cryobiology* **8,** 453–464.

Atkinson, R. G. (1953). *Can. J. Bot.* **32,** 542–547.

Atkinson, R. G. (1954). *Can. J. Bot.* **32,** 673–678.

Barnett, J. A., Payne, R. W., and Yarrow, D. (1983). "Yeasts, Characteristics and Identification." Cambridge Univ. Press, Cambridge, UK.

Bennett, J. W., and Lasure, L. L., eds. (1985). "Gene Manipulations in Fungi." Academic Press, Orlando, FL.

Boeswinkel, H. J. (1976). *Trans. Br. Mycol. Soc.* **66,** 183–185.

Booth, C. (1971). "The Genus *Fusarium.*" Commonwealth Mycological Institute, Kew, UK.

Bridge, P. D. (1985). *J. Gen. Microbiol.* **131,** 1887–1895.

Bridge, P. D., and Hawksworth, D. L. (1984). *Microbiol. Sci.* **1,** 232–234.

Bridge, P. D., and Hawksworth, D. L. (1990). *Genet. Eng. Biotechnol.* **10,** 9–11.

Bromfield, K. R., and Schmitt, C. G. (1967). *Phytopathology* **57,** 1133–1134.

Buell, C. B., and Weston, W. H. (1947). *Am. J. Bot.* **34,** 555–561.

Butterfield, W., Jong, S. C., and Alexander, M. J. (1974). *Can. J. Microbiol.* **20,** 1665–1673.

Calcott, P. H. (1978). "Freezing and Thawing Microbes." Meadowfield Press, Shildon, UK.

Castellani, A. (1939). *J. Trop. Med. Hyg.* **42,** 225–226.

Castellani, A. (1967). *J. Trop. Med. Hyg.* **70,** 181–184.

Clark, G., and Dick, M. W. (1974). *Trans. Br. Mycol. Soc.* **63,** 611–612.

Dahman, H., Staub, T., and Schwinn, F. (1983). *Phytopathology* **73,** 241–246.

Ellis, J. J. (1979). *Mycologia* **71,** 1072–1075.

Figueiredo, M. B., and Pimentel, C. P. V. (1975). *Summa Phytopathol.* **1,** 299–302.

Fincham, J. R. S., and Ravetz, J. R. (1991). "Genetically Engineered Organisms Benefits and Risks." Open University Press, Milton Keynes, UK.

Franks, F. (1981). *In* "Effects of Low Temperature on Biological Membranes" (G. J. Morris and A. Clarke, eds.), pp. 3–19. Cambridge Univ. Press, Cambridge, UK.

Frisvad, J. C. (1989). *Bot. J. Linn. Soc.* **99,** 81–95.

Gale, A. W., Schmitt, C. G., and Bromfield, K. R. (1975). *Phytopathology* **65,** 828–829.

Gams, W., van der Aa, H. A., van der Plaats-Niterink, A. J., Samson, R. A., and Stalpers, J. A. (1987). "CBS Course of Mycology." Centraalbureau voor Schimmelcultures Baarn, Netherlands.

Gentles, J. C., and Scott, E. (1979). *Sabouraudia* **17,** 415–418.

Goldie-Smith, E. K. (1956). *J. Elisha Mitchell Sci. Soc.* **72,** 158–166.

Gordon, W. L. (1952). *Can. J. Bot.* **30,** 209–251.

Grout, B. W. W., and Morris, G. J., eds. (1987). "The Effects of Low Temperatures on Biological Systems." Edward Arnold, London.

Hawksworth, D. L. (1991). *Mycol. Res.* **95,** 641–655.

Hawksworth, D. L., and Kirsop, B. E., eds. (1988). "Living Resources for Biotechnology: Filamentous Fungi." Cambridge Univ. Press, Cambridge, UK.

Hawksworth, D. L., Sutton, B. C., and Ainsworth, G. C. (1983). "Ainsworth and Bisby's Dictionary of Fungi," 7th ed. Commonwealth Mycological Institute, Kew, UK.

Heckly, R. J. (1978). *Adv. Appl. Microbiol.* **24,** 1–53.

Holden, A., and Smith, D. (1992). *Mycol. Res.* **96,** 473–476.

Hwang, S.-W. (1960). *Mycologia* **52**, 527–529.

Hwang, S.-W. (1966). *Appl. Microbiol.* **14**, 784–788.

Hwang, S.-W., and Howells, A. (1968). *Mycologia* **60**, 622–626.

Hwang, S.-W., Kwolek, W. F., and Haynes, W. C. (1976). *Mycologia* **68**, 377–387.

International Mycological Institute (1991). "Index of Fungi." CAB International, Kew, UK.

Jong, S. C., Levy, A., and Stevenson, R. W. (1984). *In* "Proceedings of the Fourth International Conference on Culture Collections" (M. Kocur and E. daSilva, eds.), pp. 125–136. World Federation for Culture Collections, London.

Kirsop, B. E., and Doyle, A., eds. (1991). "Maintenance of Microorganisms and Cultured Cells," 2nd ed. Academic Press, London.

Kirsop, B. E., and Kurtzman, C. P., eds. (1988). "Living Resources for Biotechnology: Yeasts." Cambridge Univ. Press, Cambridge, UK.

Kreger-van Rij, N. J. W., ed. (1984). "The Yeast: A Taxonomic Study," 3rd ed. Elsevier, Amsterdam.

Lapage, S. P., and Redway, K. F. (1974). "Public Health Laboratory Service," Monogr. Ser. No. 7. H. M. Stationary Office, London.

Leach, C. M. (1962). *Can. J. Bot.* **40**, 151–161.

Li Zhong-qing and Yuan Wei (1984). *Acta Mycol. Sin.* **3**, 1–3.

Lodder, J., and Krefer-van Rij, N. J. W., eds. "The Yeasts: A Taxonomic Study." North-Holland Publ., Amsterdam.

Long, R. A., Woods, J. M., and Schmitt, G. C. (1978). *Plant Dis. Rep.* **62**, 479–481.

Malik, K. A. (1991). *In* "Maintenance of Microorganisms and Culture Cells" (B. E. Kirsop and A. Doyle, eds.), pp. 81–100. Academic Press, London.

Marx, D. H., and Daniel, W. J. (1976). *Can. J. Microbiol.* **22**, 338–341.

Morris, G. J. (1981). "Cryopreservation: An Introduction to Cryopreservation in Culture Collections." Culture Centre of Algae and Protozoa, Cumbria, UK.

Morris, G. J., Smith, D., and Coulson, G. E. (1988). *J. Gen. Microbiol.* **134**, 2897–2906.

Ogata, W. N. (1962). *Neurospora Newsl.* **1**, 13.

Onions, A. H. S. (1971). *Methods Microbiol.* **4**, 113–151.

Onions, A. H. S. (1977). *In* "Proceedings of the Second International Conference on Culture Collections," (Pestona de Castro, V. B. D. Da Silva, and W. W. Leveritt, eds.) pp. 104–113. World Federation for Culture Collections, University of Queensland, Brisbane, Australia.

Onions, A. H. S. (1983). *In* "The Filamentous Fungi. 4. Fungal Technology" (J. E. Smith, D. R. Berry, and B. Kristiansen, eds.), pp. 375–390. Edward Arnold, London.

Onions, A. H. S., and Smith, D. (1984). "Critical Problems of Culture Collections," Symp. IMC3, pp. 41–47. Institute of Fermentation, Osaka, Japan.

Onions, A. H. S., Allsopp, D., and Eggins, H. O. W., eds. (1981). "Smith's Introduction to Industrial Mycology." Edward Arnold, London.

Paterson, R. R. M. (1986). *J. Chromatogr.* **368**, 249–264.

Pearson, B. M., Jackman, P. J. H., Painting, K. A., and Morris, G. J. (1990). *Cryo-Letters* **11**, 205–210.

Perkins, D. D. (1962). *Can. J. Microbiol.* **8**, 591–594.

Raper, K. B., and Alexander, D. F. (1945). *Mycologia* **37**, 499–525.

Reinecke, P., and Fokkema, N. J. (1979). *Trans. Br. Mycol. Soc.* **72**, 329–331.

Reischer, H. S. (1949). *Mycologia* **41**, 177–179.

Rey, L. R. (1977). *In* "Development in Biological Standardisation. International Symposium of Freeze Drying of Biological Products" (V. J. Cabasso and R. H. Regamey, eds.), pp. 19–27. S. Karger, Basel.

Samson, R. A., and van Reenen-Hoekstra, E. S. (1988). "Introduction to Food-borne Fungi" Centraalbureau voor Schimmelcultures, Baarn, Netherlands.

Shearer, B. L., Zeyen, R. J., and Ooka, J. J. (1974). *Phytopathology* **64**, 163–167.

Smith, D. (1982). *Trans. Br. Mycol. Soc.* **79**, 415–421.

Smith, D. (1983a). *Trans. Br. Mycol. Soc.* **80**, 333–337.

Smith, D. (1983b). *Trans. Br. Mycol. Soc.* **80**, 360–363.

Smith, D. (1986). "Ph.D. Thesis," London University.

Smith, D. (1988). *In* "Living Resources for Biotechnology: Filamentous Fungi" (D. L. Hawksworth and B. E. Kirsop, eds.), pp. 75–99. Cambridge Univ. Press, Cambridge, UK.

Smith, D. (1991). *In* "Maintenance of Microorganisms and Cultured Cells" (B. E. Kirsop and A. Doyle, eds.), 2nd ed., pp. 133–159. Academic Press, London.

Smith, D., and Onions, A. H. S. (1983a). "The Preservation and Maintenance of Living Fungi." Commonwealth Mycological Institute, Kew, UK.

Smith, D., and Onions, A. H. S. (1983b). *Trans. Br. Mycol. Soc.* **81**, 535–540.

Smith, D., Tintigner, N., Hennebert, G. L., de Bievre, C., Roquebert, M. F., and Stalpers, J. A. (1990a). *In* "Biochemistry R and D in the EC 1, Catalogue of BAP Achievements" (A. Vassarotti and E. Magnien, eds.), pp. 115–117. Elsevier, Paris.

Smith, D., Tintigner, N., Hennebert, G. L., de Bievre, C., Roquebert, M. F., and Stalpers, J. A. (1990b). *In* "Biotechnology in the EC II. Detailed Final Report of BAP Contractors" (A. Vassarotti and E. Magnien, eds.), p. 89. Elsevier, Paris.

Snyder, W. C., and Hansen, H. N. (1946). *Mycologia* **59**, 600–69.

Tetsuka, Y., and Katsuya, K. (1983). *Ann. Phytopathol. Soc. Jpn.* **49**, 731–735.

Tommerup, I. C., and Kidby, D. K. (1979). *Appl. Environ. Microbiol.* **37**, 831–835.

Trinci, A. P. J. (1992). *Mycol. Res.* **96**, 1–13.

van Ardsell, J. N., Kwok, S., Schweickart, V. L., Ladner, M. B., Gelfand, D. H., and Innis, M. A. (1987). *Bio/Technology* **5**, 60–64.

von Arx, J. A., and Schipper, M. A. A. (1978). *Adv. Appl. Microbiol.* **24**, 215–236.

Webster, J., and Davey, R. A. (1976). *Trans. Br. Mycol. Soc.* **67**, 543–544.

Xie, Y.-M., and Cong, S. H. (1987). *Acta Mycol. Sin.* **6**, 51–57.

Yarlett, N. C., Yarlett, N., Orpin, C. G., and Lloyd (1986). *Lett. Appl. Microbiol.* **3**, 1–3.

CHAPTER 7

Protozoa

Ellen M. Simon

Thomas A. Nerad

Background

Introduction

Protozoa have been defined as "one-celled or colonial (polycellular) organism[s] possessing all of the elegantly coordinated processes of life, usually (though not exclusively) animal like within the boundaries of a single cell" (Anderson, 1988). The protozoa are significant components of the biota of diverse ecosystems; they exist in soils from polar regions to hot spring runoffs, in marine, estuary, or fresh water, at all altitudes, in sewage treatment systems, and in droplets trapped in flowers and banana plants. Their associations with other organisms include symbiosis, commensalism, and parasitism—for example, the obligate parasitism of cellulose-digesting flagellates in termites (Breznak and Brune, 1994) and wood roaches and of ciliates in the digestive systems of ruminants, or as the etiological agents of widespread human and animal diseases (Nisbet, 1984).

Most protozoa have attributes characteristic of "higher" animals, e.g., motility, in at least some phase of their life cycle, and acquisition of nutrients by ingestion (phagocytosis or pinocytosis). Some, however, are capable of photosynthesis, active transport, or diffusion, and there are examples among the euglenoids (Lee *et al.*, 1985) and dinoflagellates (Bochstahler and Coats, 1993) of organisms capable of living both autotrophically and heterotrophically. In addition to these methods for obtaining nutrients, some ciliates have the means to adopt a cannibalistic life style. Most species of *Tetrahymena* ingest bacteria, absorb nutrients from axenic media (containing no other organisms); other species, by transforming into macrostome cells, can also ingest other tetrahymenas. Free-living flagellates and ciliates are crucial components of terrestrial and aquatic food webs, where they consume tiny primary producers and are in turn consumed by small metazoa, or they contribute to mineralization of essential nutrients (Fenchel, 1993; Hall *et al.*, 1993; Vickerman, 1992). As part of the complex microbial communities in sewage treatment they contribute to the reduction of organic matter and mineralization (Aitken *et al.*, 1992). Some species that enclose themselves in shells, or tests, containing silicon or calcium, after death of the cell, may contribute to the formation of limestone (Lee *et al.*, 1985).

In contrast to the helpful activities mentioned above, there are pathogenic protozoa, usually considered in the province of parasitology, that cause much suffering, death, and economic devastation (see Levine, 1985). Members of the Apicomplexa (Sporozoa) cause malaria (*Plasmodium*), which still kills a million people a year. *Babesia* and *Theileria* produce malaria-like diseases in animals by infecting erythrocytes. Coccidia of several genera cause intestinal diseases in domestic and wild animals. Toxoplasmosis is generally a feline disease that is easily transmitted to humans, in which it frequently causes congenital anomalies and has been implicated as the most common cause of focal central nervous system infection in AIDS patients (Joiner and Dubremetz, 1993). A related genus, *Sarcocystis*, infects herbivores and ducks. Several species of microsporidia also adversely affect AIDS patients (Asmuth *et al.*, 1994). Ishii *et al.* (1994) reviewed evidence for a connection between parasites, especially *Schistosoma*, and cancer. The sporozoans have complex life cycles, usually involving two different hosts.

Among diseases caused by the flagellated protozoa are the sexually transmitted trichomoniasis (Lee, 1985a), water-transmitted giardiasis (Lee, 1985b), insect-transmitted kala-azar (*Leishmania*), Chaga's disease, African sleeping sickness, and nagana, the last three caused by trypanosomes (Lee and Hutner, 1985). Nagana precludes the raising of cattle in much of central Africa, probably because they lack a trypanocidal protein present in the African buffalo (Reduth *et al.*, 1994). Other trypanosomes are plant parasites that affect latex plants, coffee plants, and oil and coconut palm trees. Several amoebae are responsible for human suffering—e.g., amoebic dysentery (*Entamoeba*), frequently fatal meningitis (*Naegleria*), and eye infections (*Acanthamoeba*) for users of soft contact lenses (Bard and Lambrozo, 1992). Some amoebae act as reservoirs for *Legionella*, a human respiratory pathogen (Srikanth and Berk, 1993). The ciliates include at least one significant mammalian pathogen, *Balatidium coli,* an inhabitant of mammalian intestines that can infect humans (Lee *et al.*, 1985), and the economically important fish pathogen, *Ichthyophthirius multifiliis* (Dickerson *et al.*, 1989).

As experimental organisms the protozoa have made significant contributions to our understanding of basic biological phenomena. Earlier work has been summarized in Levandowsky and Hutner (1979–1981). Three major recent discoveries are (1) the structure and function of telomeres, which provided the explanation for the stability of the ends of chromosomes (Blackburn, 1986); (2) the first demonstration of self-splicing RNA, which derailed the dogma that proteins were required for all "enzymatic" reactions (Cech, 1986); and (3) the first exceptions to the "universal" code for the translation of mRNA into protein—TAA, usually a "stop" signal, codes for glutamine in several genera (also TAG in *Paramecium*); in *Euplotes*, UGA, which usually has no complementary tRNA, codes for cysteine (Meyer *et al.*, 1991). These advances were made possible by the unique attributes of the ciliates. Most are large cells with the usual eukaryotic organelles, the unique possession of germinal and somatic nuclei in the same cytoplasm, and the capacity to produce large populations, often in defined media. The 9000- to 10,000-fold selective amplification of the rRNA gene in the *Tetrahymena* macronucleus contributed significantly to the discoveries of telomeres and RNA splicing.

Studies of the development of somatic nuclei in ciliates and of the variable surface antigens of the flagellated trypanosomes have also contributed to our understanding of DNA rearrangements (Prescott, 1992; Steele *et al.*, 1994) and other aspects of development (Yanagi, 1992).

At least two of the ciliates, *Stylonichia* (Machemer, 1987) and *Paramecium* (Kung *et al.*, 1992), have increased our understanding of membrane potentials, ion channels, electrical excitation and behavior, and signal transduction (Ginsburg *et al.*, 1995; Luporini *et al.*, 1995; Van Houten, 1994), phenomena related to neuroscience. [For additional examples of the contributions of the ciliates to molecular biology until the mid-1980s see Gall (1986).] *Tetrahymena* cells have opioid receptors that could provide a model to help us understand how opioids affect nerve cells (Chiesa *et al.*, 1993; Renaud *et al.*, 1995). Rubin (1994) credits early genetic studies on ciliates with the first description of nongenetic heredity, an observation expanded by the technique of microsurgery and by the study of ciliate mutants that reveal surface pattern phenomena and cytoplasmic positional information that could apply to embryology (Fleury and Laurent, 1994; Frankel, 1989). The genus *Naegleria* offers a unique model in its rapid, synchronous change from amoeboid to flagellated cells (Fulton, 1993). The slime molds *Dictyostelium* (Raper, 1984) and *Physarum* (Anderson, 1992), claimed by both protozoologists and mycologists (see Figure 1), offer opportunities to ob-

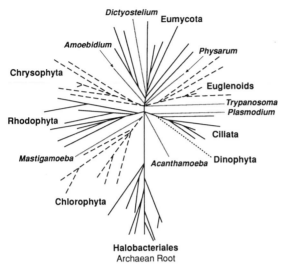

Figure 1. A tree summarizing the eukaryotic radiation based on published sequences of the 5S rRNAs (Erdmann and Wolters, 1988) analyzed with the Phylogen 1.0 Program (Nanney *et al.*, 1995). It is rooted in the sequences of 10 strains of the Halobacteriales. The tip of each branch represents a particular 5S sequence. All bifurcations represent calculated ancestor sequences. The six major protist groups are shown to have arisen explosively during an evolutionarily brief period. Each group forms a coherent branch, which correlates very well with long-recognized "natural" groupings. The single available representatives of two of the major pathogenic genera are included. *Trypanosoma brucei* is connected to the base of the euglenoid branch and *Plasmodium berghei* is connected to the base of the ciliate branch. The five amoeboid sarcodinids, including *Physarum* and *Dictyostelium*, if analyzed by themselves, form a coherent branch, but when included in the complete analysis, each arises directly from the eukaryotic stem or very close to the base of a major branch. This suggests that "sarcodinids" were the earliest eukaryotes. The position of the dinoflagellate as an early emergent on the ciliate branch, not an alga, is supported by morphological characters and SSU and LSU rRNA sequences (Lee and Kugrens, 1992; Cavalier-Smith, 1993).

serve cell differentiation, intercellular communication, and pattern formation (Kessler and Levine, 1993).

Classification and Diversity

Some inhabitatns of the microbial world were well described in the late seventeenth century by Leeuwenhoek. The autotrophic, generally non-motile, organisms were classified as plants; the motile heterotrophs were classified as animals. At least two groups, the euglenoids and dinoflagellates, created problems as morphologically indistinguishable cells that could flourish as autotrophs or heterotrophs, and motility was not correlated with nutrition. More than a century ago Haeckel (1866) suggested that a third kingdom, Protista, should include all the microorganisms then known. His suggestion was not generally accepted but it was championed a century later by Whittaker (1969). Botanists and zoologists studied, named, and classified the euglenoids, dinoflagellates, and several other groups of unicellular organisms, each according to the rules of the respective codes of nomenclature and classification systems.

Margulis (1974) and Corliss (1984) have resurrected Haeckel's Kingdom Protista (or Protoctista) and a widely recognized classification was proposed by a Committee of the Society of Protozoologists (Levine et al., 1980). This scheme was generally followed in Lee et al. (1985) and the higher taxa are listed in Table 1. Corliss divided the Protista into 45 phyla based on morphological, ecological, biochemical, and molecular features (Table 2).

Cavalier-Smith (1993) has proposed an extensive restructuring of the classification of the biological world that is based principally on the nucleic acid sequences of the small subunit rRNA, the structure of plastids, and the presence or absence of starch. Corliss (1994) strongly supports his suggested six kingdoms, all of which include protists, except Animalia (Table 2).

The tree in Figure 1, based on an analysis of published 5S rRNA sequences, suggests an explosive emergence of eukaryotic protists (D. L. Nanney, C. Park, and E. M. Simon, in preparation). The well represented groups of algae, lower fungi and protozoa form separate, coherent branches generally well correlated with recognized major taxa.

The protozoa, even if we consider only the taxa remaining in Cavalier-Smith's revised Kingdom Protozoa, are a fantastically diverse group. Symmetry may be radial, bilateral, or neither. Their surface structures vary from membrane-bound naked protoplasm (some amoebas), to the extremely complex cortex of the ciliates, to dozens of different types of shells or tests. In length, or diameter, they range from ~2 μm to >1 cm. Members of different taxa have become adapted to essentially all types of niches in nature that will support life. A small sample of water, soil, or even rumen contents may yield a mixture of species or genera. Many taxa produce cysts that are resistant to adverse environmental conditions and may remain viable for years.

Much diversity of a different kind exists within individual morphologically defined species. More than 50 years ago Sonneborn (1939) demonstrated that the "species" *Paramecium aurelia* was a complex of biologically isolated populations for which he coined the term syngens. Three decades later, when the syngens could be identified by isozyme electrophoretic

TABLE 1
Higher Taxa in the Classification of the Subkingdom Protozoa[a]

Phylum Subphylum Superclass Class	Orders (groups)	Phylum Subphylum Superclass Class	Orders (groups)
Sarcomastigophora		**Apicomplexa**	11
Mastigophora		**Microspora**	
Phytomastigophorea	10	Metchnikovellidea	
Zoomastigophorea	11	Microsporididea	2
Opalinata		**Myxozoa**	
Opalinatea	1	Myxosporea	2
Sarcodina		Actinosporea	1
Rhizopodea		**Ciliophora**	
Lobosea	5	Postciliodesmatophora	
Acarpomyxea	2	Karyorelictea	4
Mycetozoea	9	Spirotrichea	10
Filosea	2	Rhabdophora	
Granuloreticulosea	3	Prostomatea	2
Actinopodea		Litostomatea	5
Acantharea	4	Cyrtophora	
Polycystinea	5	Phyllopharyngea	7
Phaeodarea	7	Nassophorea	6
Heliozoea	6	Oligohymenophorea	9
Labyrinthomorpha		Colpodea	4
Labyrinthulea	1		

[a] From Lee *et al.* (1985).

polymorphisms, he assigned binomial species names to them. That this is far from a unique situation has been shown at least for many ciliated protozoa (reviewed by Dini and Nyberg, 1993); 20 of 24 morphological species examined included from 2 to 16 biologically isolated sibling species. Among nonciliates the same phenomenon has been reported for a slime mold, i.e., *Dictyostelium* (Raper, 1984), and for some green algae, i.e., *Chlamydomonas* (Harris, 1989) and *Pandorina* (A. Coleman and R. N. Preparata, personal communication). The practice of defining genera or species by the animal or plant host from which a protist was isolated is not well supported. For an example of the taxonomic insufficiency of host origin and morphology, see Teixeira *et al.* (1995). On the other hand, Stringer *et al.* (1993) advocate caution in the use of a rat model for *Pneumocystis carinii* infections; in the four different types of molecular tests, rat and human isolates were distinct.

The differentiation of sibling species may be more complicated in genera other than *Paramecium*. The published analysis of the *Tetrahymena pyriformis* complex (Meyer and Nanney, 1987) suggested that the asexual (amicronucleate) and the 15 reproductively isolated sexual sibling species were differentiated by their phenotypes in a limited number of enzyme systems. As more strains were examined by Meyer or either of us (unpublished), additional intraspecific polymorphism destroyed some of the presumed "unique" phenotypes. Isozyme electrophoresis has been widely used with nonciliates in epidemiology and for characterization of free-living species (Guerrini *et al.*, 1992).

The importance of publishing the source of strains used and their *complete* strain identifications cannot be overemphasized. *Tetrahymena pyriformis*—if it is *sensu stricto,* is it GL, H, W, ST, etc.? If it is *sensu lato,* it could be any one of ~20 morphologically indistinguishable species whose molecules have been diverging for millions of years. Another example, *Tetrahymena pigmentosa,* strain 6UM, drastically narrows the possibilities, but it is incomplete because there are 6UM strains with different mating types and different surface antigens.

Economic Uses

Although the protozoa have not contributed directly to satisfying human appetites as have the bacteria, yeasts, and molds in the production of dairy products, bread, beer, wine, spirits, and many oriental foods, or in direct consumption as have some algae, some protozoa have been utilized in ways that benefit evolutionarily higher forms of life.

AS ASSAY AND INDICATOR ORGANISMS

In the estimation and monitoring of water quality, protozoa have made significant contributions. In unpolluted water there is a very diverse community. As pollution increases, the community is reduced until very few species are present. Protozoan populations can be used for detection and quantitative characterization of environmental factors present as a result of natural or human influence (Fenchel, 1987; Foissner, 1987). Sensitive assays for heavy metals may use *Colpidium* (Sekkat *et al.*, 1992), *Colpoda* (Forge *et al.*, 1993), or *Tetrahymena* (Noever *et al.*, 1992). However, heavy metals may have a detrimental effect on *Euglena* when it is used as an indicator of pollution (Stallwitz and Hader, 1993). Assays for hydrocarbons and an insecticide (Komala, 1993) have used *Paramecium.* A laboratory evaluation of the effects of lead on marine protozoan communities showed drastic

TABLE 2
Summary of Methodology and Holdings at the American Type Culture Collection

Assemblage phylum[a]	Cultivation method[b]		Preservation method[b]			ATCC holdings[c]				Representative genera
	Xenic	Axenic	Drying	Freeze-drying	Freezing	No. of genera	No. of species	No. of strains	Kingdom[d]	
Rhizopods										
Acrasia[e]	−	−	−	−	−	1	1	1	Protozoa	*Fonticula*
Amoebozoa	+	+	+	+	+	35	79	188	Protozoa	*Acanthamoeba, Naegleria, Entamoeba*
Eumycotozoa[e]	+	+	−	+	+	10	58	282	Protozoa	*Dictyostelium, Physarum*
Granuloreticulosa	+	−	−	−	+	1	1	1	Protozoa	*Allogromia*
Karyoblastea	+	+	−	−	+	2	2	4	Archezoa	*Mastigamoeba, Phreatamoeba*
Plasmodiophorea	−	−	−	−	−	0	0	0	Protozoa	—
Xenophyophora	−	−	−	−	−	0	0	0	Protozoa	—
Mastigomycetes										
Chytridiomycota[e]	−	−	−	−	−	14	33	69	Fungi	*Allomyces, Chytrium*
Hyphochytridiomycota[e]	−	−	−	−	−	1	1	4	Fungi	*Hypochytrium*
Oomycota[e]	−	−	−	−	−	15	191	1222	Fungi	*Achlya, Phytophthora, Pythium*
Chlorobionts										
Charophyta	−	−	−	−	−	0	0	0	Plantae	—
Chlorophyta	+	+	+	+	+	35	79	133	Plantae	*Chlamydomonas, Chlorella*
Conjugaphyta[f]	−	−	−	−	−	1	1	1	Plantae	*Mesotaenium*

(*Continues*)

									Kingdom	Genera
Glaucophyta	—	—	—	—	—	—	0	0	0	Plantae	—
Prasinophyta	+	+	—	—	+	+	1	1	3	Plantae	*Tetraselmis*
Euglenozoa											
Euglenophyta	+	+	—	—	+	+	8	16	20	Protozoa	*Astasia, Euglena*
Kinetoplastidea	+	+	+	+	+	+	14	82	172	Protozoa	*Bodo, Crithidia, Leishmania*
Pseudociliata	+	—	—	—	+	+	1	1	1	Protozoa	*Stephanopogon*
Rhodophytes											
Rhodophyta[f]	—	—	—	—	+	+	1	1	2	Plantae	*Porphyridium*
Cryptomonads											
Cryptophyta	+	+	—	—	+	+	2	2	2	Chromista	*Chroomonas, Goniomonas*
Choanoflagellata											
Choanoflagellata	—	—	—	—	—	+	7	7	7	Protozoa	*Monosiga, Salpingoeca Diaphanoeca*
Chromobionts											
Bacillariophyta[f]	—	—	—	—	—	+	1	1	1	Chromista	*Amphiprora*
Bicosoecidea	+	—	—	—	+	+	3	4	4	Chromista	*Bicosoeca, Cafeteria*
Chrysophyta	+	+	—	—	+	+	5	5	5	Chromista	*Ochromonas, Poteriochromonas*
Eustigmatophyta[f]	—	—	—	—	—	+	4	4	4	Chromista	*Monodus, Polyedriella*
Haptophyta	+	—	—	—	+	+	1	1	1	Chromista	*Pavlova*
Heterochloridia	—	—	—	—	—	—	0	0	0	Chromista	—
Phaeophyta	—	+	—	—	+	—	0	0	0	Chromista	—
Proteromonadea	+	—	+	—	+	+	1	1	1	Protozoa	*Proteromonas*
Raphidophycea	—	—	—	—	—	—	0	0	0	Chromista	—
Xanthophyta[f]	—	—	—	—	+	+	5	7	7	Chromista	*Botrydium, Tribonema*

142

TABLE 2 (*Continued*)

Assemblage phylum[a]	Cultivation method[b]		Preservation method[b]			ATCC holdings[c]			Kingdom[d]	Representative genera
	Xenic	Axenic	Drying	Freeze-drying	Freezing	No. of genera	No. of species	No. of strains		
Labyrinthomorphs										
Labyrinthulea	+	−	−	−	+	0	0	0	Chromista	*Labyrinthula*
Thraustochytriaceae[e]	−	−	−	−	−	3	10	18	Chromista	*Thraustochytrium*
Polymastigotes										
Metamonadea	+	+	−	−	+	3	7	13	Archezoa	*Giardia, Hexamita*
Parabasalia	+	+	−	−	+	10	16	82	Protozoa	*Dientamoeba, Trichomonas*
Paraflagellates										
Opalinata	−	−	−	−	−	0	0	0	Protozoa	—
Actinopods										
Acantharia	−	−	−	−	−	0	0	0	Protozoa	—
Heliozoa	−	−	−	−	−	0	0	0	Protozoa	—
Phaeodaria	−	−	−	−	−	0	0	0	Protozoa	—
Polycistina	−	−	−	−	−	0	0	0	Protozoa	—
Taxopoda	−	−	−	−	−	0	0	0	Protozoa	—
Dinoflagellates										
Acritarcha	−	−	−	−	−	0	0	0	Protozoa	—
Ebriidea	−	−	−	−	−	0	0	0	Protozoa	—
Ellobiophyceae	−	−	−	−	−	0	0	0	Protozoa	—
Peridinea	+	+	−	−	+	2	2	45	Protozoa	*Amphidium, Crypthecodinium*
Syndinea	−	−	−	−	−	0	0	0	Protozoa	—

Ciliates									
Ciliophora	+	+	+	−	30	88	418	Protozoa	*Paramecium, Tetrahymena*
Sporozoa									
Perkinsida	−	−	−	−	1	1	1	Protozoa	*Perkinsus*
Sporozoa	−	+	−	+	4	21	81	Protozoa	*Babesia, Toxoplasma*
Microsporidia									
Microsporidia	−	+	−	+	2	4	4	Archezoa	*Nosema, Pleistophora*
Haplosporidia									
Haplosporidia	−	−	−	−	0	0	0	Protozoa	—
Myxosporidia									
Actinomyxidea	−	−	−	−	0	0	0	Protozoa	—
Marteiliidea	−	−	−	−	0	0	0	Protozoa	—
Myxosporidia	−	−	−	−	0	0	0	Protozoa	—
Paramyxidea	−	−	−	−	0	0	0	Protozoa	—
Taxa of uncertain affinities[g]	+	+	+	+	13	18	19	Protozoa, Chromista	*Amastigomonas, Heteromita*

[a] Phyla Protista *sensus* Corliss (1984).
[b] Cultivation and preservation protocols published in "Protocols in Protozoology" (Lee and Soldo, 1992).
[c] Current protist holdings at the ATCC.
[d] Classification scheme of Cavalier-Smith (1993).
[e] Phyla listed in the ATCC Catalogue of Protists (Nerad, 1993) but maintained in the Mycology and Botany Department.
[f] Preservation methods given in Simione and Brown (1991).
[g] Includes largely heterotrophic flagellates of uncertain taxonomic affinities. Cavalier-Smith (1993) places some of these taxa in the kingdom Protozoa and others in the kingdom Chromista.

reductions in the number of taxa and protozoan biomass (Fernandez-Leborans and Novillo, 1992). Sugiura (1992) has screened for several toxic chemicals with a microcosm of prokaryotic and eukaryotic species. Madoni (1994) has evaluated biological performance in activated sludge plants by analyses of microfauna.

Various species of *Amphidinium, Crithidia, Crypthecodinium, Euglena, Ochromonas, Selenastrum,* and *Tetrahymena* are used in assays for some vitamins, amino acids, etc.; *Acanthamoeba, Chlorella, Crithidia, Dunaliella, Giardia, Trichomonas,* and *Tritrichomonas* produce several enzymes or glycerol; a vaccine against *Leishmania tropica* is available; *Chlorella saccharophila* is resistant to acid or NaCl; drug-sensitive and/or -resistant strains of *Acanthamoeba, Plasmodium,* and *Trichomonas* are available; *Nosema locustae* is a biocontrol agent for grasshoppers and *Prototheca zopfii* can degrade petroleum. There are also many endosymbiont-bearing strains, mostly paramecia, and several dozen genetically characterized strains of *Chlamydomonas, Paramecium,* and *Tetrahymena.* The information in this paragraph has been taken from the Catalogue of Protists, The American Type Culture Collection (Nerad, 1993). The specific strains are listed in the "Special Applications" section (pp. 77–79), and relevant references, too many to list here, are included in the strain descriptions.

Unlike plants and several categories of microorganisms, the nonpathogenic protozoa have not been the subject of extensive searches for pharmacologically active natural products of interest. This situation may be changing. Tetrahymenol, a membrane component, may be a specific marker for marine ciliates (Harvey and McManus, 1991); D- and L-pterins have been found in ciliates and flagellates (Klein and Groliere, 1993).

More possibilities for biological control include that of zebra mussels by some ciliates and bacteria (Toews *et al.,* 1993), of *Acanthamoeba* by *Pseudomonas* (Qureshi *et al.,* 1993), of *Naegleria* by a *Bacillus* (Cordovilla *et al.,* 1993), and of gypsy moths by *Nosema* (Novotny and Weiser, 1993). Pidherney *et al.* (1993) described tumoricidal properties of an amoeba.

THE SEARCH FOR IMPROVED AGENTS FOR PROPHYLAXIS OR TREATMENT OF PROTOZOAN DISEASES

The successes obtained with prophylactic vaccines in eliminating smallpox or in reducing the incidence of other viral and some bacterial diseases have proved to be much more elusive for the diseases caused by protozoa. Likewise, efforts to find effective therapeutic agents are frustrating. A major factor contributing to this problem is that, as eukaryotic cells, the organelles and metabolic processes of the protozoa are closely related to those of higher animals. Except when drug resistance has developed, the sulfa drugs and antibiotics have had a profound effect on diseases caused by prokaryotic bacteria; this is because these drugs attack structures or processes in prokaryotic cells that are quite different from their analogs in eukaryotic cells.

Much effort has been directed toward the identification of molecular components of protozoan etiological agents that may provide potential components of vaccines or targets for chemotherapy. Related topics for *Plasmodium, Theileria, Leishmania, Trypanosoma, Trichomonas,* and *Giardia* were reviewed in Turner and Arnot (1988). A few additional research papers concern the amoebae (Das *et al.,* 1993; Hadas and Mazur, 1993; Horstmann *et al.,* 1992; John and Howard, 1993; Mukherjee *et al.,* 1993; Toney and Marciano-Cabral, 1994), *Plasmodium* (Curley *et al.,* 1994; Kaidoh *et al.,* 1995; Sam-Yellowe, 1992; Sam-Yellowe *et al.,* 1995), *Babesia* (Dalrymple, 1993; Valentin *et al.,* 1993), *Toxoplasma* (Saffer *et al.,* 1992; Moulin *et al.,* 1993; Sibley *et al.,* 1993), *Trichomonas* (Gold, 1993), the trypanosomatids (Bates *et al.,* 1994; Marche *et al.,* 1993; Osuna *et al.,* 1984; Souto-Padron *et al.,* 1994; Yang *et al.,* 1994), and host defenses (Hall and Joiner, 1993; Shakibaei and Frevert, 1992). The proceedings of a symposium "Biochemical Protozoology as a Basis for Drug Design" (Coombs and North, 1991) and other reports suggest possible targets in *Trypanosoma* (Wheeler-Alm and Shapiro, 1992), *Acanthamoeba* (Mehlotra and Shukla, 1993), and *Toxoplasma* (Iltzsch, 1993). New drugs are being evaluated for *Plasmodium* (Basco and leBras, 1994; Presber, 1994) and *Giardia* (Cooperstock *et al.,* 1992) and *Pneumocystis* (Combly and Sterling, 1994).

Molecular techniques for diagnosis or epidemiological investigations include recombinant proteins (Lotter *et al.,* 1992), immunological tests (Garcia and Bruckner, 1993; Hague *et al.,* 1993), isozyme analysis (Guerrini *et al.,* 1992), and DNA or rRNA analyses (Alves *et al.,* 1994; Dirie *et al.,* 1993; Sparagano, 1993; Uliana *et al.,* 1994; Van Eys *et al.,* 1992). Visvesvara (1993) reviewed information on the epidemiology of amoebic infections and on laboratory diagnosis of microsporidiosis. Many abstracts of presentations at the Ninth International Congress of Protozoology (Berlin, July, 1993) dealt with research on pathogenic protozoa. The Society of Protozoologists (1994) sponsored workshops on *Pneumocystis, Toxoplasma* microsporidia, and *Cryptosporidium* and published 95 brief communications by active researchers.

Walzer (1993) has edited a revised and greatly expanded volume on *Pneumocystis carinii* pneumonia, first published in 1984. Other books concerning the pathogenic protozoa include Avila and Harris (1992), Garcia and Bruckner (1993), Hyde (1993), Kierszenbaum (1994), Maizels *et al.* (1992), and several volumes edited by Kreier (1991–1994).

Methods for Preservation

Collection, Isolation, and Cultivation

Diverse protozoa inhabit moist soil or any natural body of water. Some live in commensal, symbiotic, or parasitic relationships with other protists, animals, and a few plants. Their life styles and nutritional requirements preclude the development of any generally applicable methods for collection, isolation, or cultivation. Relevant references are the "Illustrated Guide to the Protozoa" (Lee *et al.*, 1985), which contains descriptions of thousands of genera belonging to all groups of protozoa, and the "Protocols in Protozoology" (Lee and Soldo, 1992), a detailed methods manual with emphasis on cultivation and cryopreservation.

An attempt to correlate suggested methods with two recent classification schemes is presented in Table 2. The method of cultivation or preservation indicates procedures for Corliss's (1984) phyla. The major relevant genera are also listed as well as the location of the phylum in Cavalier-Smith's classification (1993).

General procedures for collecting, culturing, and observing free-living protozoa and some pathogens, and detailed procedures for fixing and staining cells, are given by Lee *et al.* (1985), Jerome *et al.* (1993), and Fernandez Galiano (1994). There are also extensive tables listing the types of protists and names of genera found in soil, bogs, or marshes, in fresh or polluted water, and in marine or brackish habitats (Lee *et al.*, 1985). More specific methods for collecting and quantitating ecological populations as well as media formulations and specific instructions for cultivating many protists are described in Lee and Soldo (1992), Breteler and Laan (1993), Carrick *et al.* (1992), Laybourn-Parry (1992), Lim *et al.* (1993), Lovejoy *et al.* (1993), Nerad (1993), and Starink *et al.* (1994). Some of the additional procedures that have been published include those for *Acanthamoeba* Visvesvara and Balamuth, 1975), *Babesia* (Thomford *et al.*, 1993), *Giardia Entamoeba* and/or *Trichomonas* (Diamond *et al.*, 1995; Farthing *et al.*, 1983; Lujan and Nash, 1994), *Naegleria* (Marciano-Cabral and Toney, 1994), *Plasmodium* (Rojas and Wasserman, 1993), *Leishmania* (Pan *et al.*, 1993), *Trypanosoma brucei* (Hirumi *et al.*, 1992), axenic cultivation and transmission of plant trypanosmes (Jankevicius *et al.*, 1993; Sanchez-Moreno *et al.*, 1995), *Pneumocystis* (Kaneshiro *et al.*, 1993; Lee *et al.*, 1993), *Nosema* (Yasunaga *et al.*, 1994), *Tetrahymena* (Hellung-Larsen *et al.*, 1993), *Perkinsus* (Gauthier *et al.*, 1995), *Toxoplasma* (Weiss *et al.*, 1995), naked amoebae (Kalinina and Page, 1992), flagellates and/or ciliates (Gasol, 1993; Ghilardi *et al.*, 1992; Ohman, 1992; Wheatley *et al.*, 1993), a rumen ciliate (Michalowski *et al.*, 1991), and anaerobic ciliates or flagellates (Holler *et al.*, 1994). Media formulations that are useful for cultivating the organisms in the ATCC collection are given by Nerad (1993). Reference to the appropriate media and some detailed procedures are included in the strain descriptions. Deionized, distilled, or tap water may not give satisfactory results when used in preparation of media. Natural spring water is frequently better for freshwater organisms. Pathogenic protozoa pose special problems. Some can be grown only *in vivo;* tissue culture and media with components such as serum or egg yolk are adequate for *in vitro* cultivation of others.

Samples of natural soil or water, even carcasses of some metozoa, will contain mixtures of bacteria, algae, protozoa, and metozoa. If protozoan strains are desired for experimental work, clones must be established from single cells (Sonneborn, 1950; Lee and Soldo, 1992). Detailed methods for genetic experiments with ciliate genera are given by Sonneborn (1950) for *Paramecium* and by Orias and Bruns (1976) for *Tetrahymena.* Methods are published for the production of polyclonal or monoclonal antisera (see Smith *et al.,* 1992), the purification of antigens (Doerder and Berkowitz, 1986; Preer *et al.,* 1987), isozyme electrophoresis (Meyer and Nanney, 1987), and determination of nucleic acid homologies and sequences (American Medical Television, 1993; Glick and Pasternak, 1994; Adolph, 1994).

Preservation

Many protozoans can be maintained at room temperature or lower, by periodic transfer. However, drying, freeze-drying, and cryopreservation greatly reduce the labeling errors and the probability of the loss of desired attributes through genetic change. As far as is known, viability can be maintained indefinitely at temperatures below $-130°C$, where chemical reactions are not possible. Long-term storage is also feasible at -70 to $-100°C$, but the period of viability decreases as the temperature is increased. Cysts that form part of the life cycle of some protozoa survive in the dried state; these strains and a few others can be freeze-dried. Such preparations are less stable but they can be stored less expensively and shipped more economically (Lee and Soldo, 1992).

The phyla, for some representatives of which specific procedures are given in Lee and Soldo (1992), are indicated in Table 2. In preparing cultures for cryopreservation many factors must be considered. Among the most critical are the age of the culture, the cryoprotectant used, and the procedures used for cooling and thawing. Cultures at peak density, very late log phase, or early stationary phase survive best. Dimethyl sulfoxide (DMSO) is the cryoprotective agent of choice for many genera. However, there is no standard concentration. For the amoebae, 5–7.5% is optimal for many genera, but a few survive better with 2.5, 3, 9, or 10%. The genus *Entamoeba* also requires 3% sucrose and 20% heat-inactivated bovine serum. Among the flagellates, DMSO in concentrations of 3, 5, 7.5, 9, 10, or 12% are satisfactory for most genera, but 10% DMSO plus 7.5% methanol or glycerol, or 7.5% glycerol alone, is best for other genera. The concentrations of DMSO for ciliates are 7.5, 9, 10, or 11 and some species of *Paramecium* require 1.25% bovine serum albumin in addition to 7.5% DMSO. The two genera of microsporidia at the ATCC require 50% glycerol for cryopreservation, but *Pleistophora* in insect gut can be freeze-dried. Successful protocols for the *Sporozoa* are diverse; *Eimeria* can be frozen in 7.5% DMSO; *Babesia*, in 16.7% glycerol; and *in vivo* cultures of *Plasmodium*, in 10% glycerol; but *in vitro*-grown plasmodia require 28% glycerol, 3% sorbitol, and 0.65% NaCl.

A wide variety of microorganisms and other cells can be frozen by cooling at $-1°C/min$ from 25 to $-40°C$. This rate, however, is lethal

---

Figure 2. Method for selection of new cryopreservation protocol. If one is attempting to cryopreserve a protist never before preserved, this protocol may aid in selection of a method. The first step is to select a cryoprotective agent. For most protists without chloroplasts or leukoplasts, select cryoprotective agents in the following order: DMSO, glycerol, methanol. For protists with chloroplasts or leukoplasts, choose cryoprotective agents in the following order: methanol, glycerol, DMSO. Do a toxicity test of the cryoprotective agent in the following manner: Set up a range of concentrations of the cryoprotective agent in the concentrations indicated in the figure. Do the test with concentrations in 1% increments and at the growth temperature of the organism. Determine the concentration that the cells will tolerate for 30 min with 75% or greater viability. Once this concentration is determined, do an experimental cryopreservation run using five concentrations: the concentration determined and, in addition, concentrations 1 and 2% below and above the value. Next select the cryopreservation cycle in the order indicated; there are three basic cooling cycles: (1) $-1°C/min$ from the growth temperature to $-40°C$; then plunge into liquid nitrogen; (2) $-1°C/min$ from the growth temperature to $+4°C$; $-10°C/min$ from $+4°C$ to the heat of fusion; $-1°C/min$ from the heat of fusion to $-40°C$; then plunge into liquid nitrogen; (3) $-10°C/min$ from the growth temperature to the heat of fusion; $-1°C/min$ from the heat of fusion to $-40°C$; then plunge into liquid nitrogen. The selection of the thawing cycle is often as important as the cooling cycle. It is recommended that cells be thawed using two different protocols. (1) Normal thawing cycle: Place the vial in a 35°C water bath until completely thawed. Add thawed contents to 5–10 ml of fresh medium. During the freezing cycle the cells are highly dehydrated. Osmotic shock can be a very significant factor in cell death after thawing. (2) The most gentle thawing protocol: Thaw ampoule normally; add the thawed contents to a T-25 flask. Allow the flask to remain undisturbed for 5–15 min. Aseptically double the volume by adding 0.5 ml of fresh medium. Allow the flask to remain undisturbed for an additional 5–15 min. At this time add an equal volume of fresh medium. Continue to double the volume until the final volume is 8 ml. Incubate normally. The length of the intervals between doublings of the volume may depend on the organism being studied. To further reduce osmotic shock in freshwater organisms, add 4–6% sucrose to the growth medium and thaw using the normal method and the slow dilution method. If the selection method outlined above is utilized a great many protists not currently preserved might be cryopreserved.

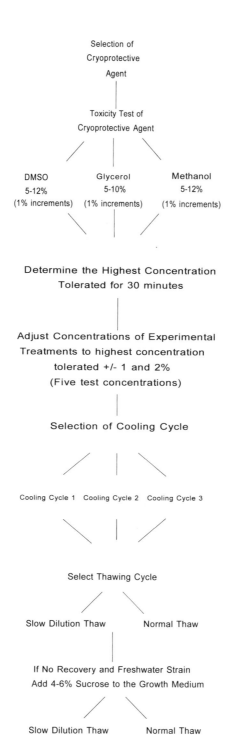

Selection of
Cryoprotective
Agent

|

Toxicity Test of
Cryoprotective Agent

DMSO
5-12%
(1% increments)

Glycerol
5-10%
(1% increments)

Methanol
5-12%
(1% increments)

Determine the Highest Concentration
Tolerated for 30 minutes

Adjust Concentrations of Experimental
Treatments to highest concentration
tolerated +/- 1 and 2%
(Five test concentrations)

Selection of Cooling Cycle

Cooling Cycle 1   Cooling Cycle 2   Cooling Cycle 3

Select Thawing Cycle

Slow Dilution Thaw   Normal Thaw

If No Recovery and Freshwater Strain
Add 4-6% Sucrose to the Growth Medium

Slow Dilution Thaw   Normal Thaw

to some species of *Tetrahymena* and other genera. A cooling rate of −10°C/min to the heat of fusion, or at least from +4°C to the heat of fusion, often solves this problem.

Cryopreserved cultures must be thawed rapidly, usually in a 35°C waterbath, and any DMSO must be diluted to nontoxic levels as soon as possible. The cells have been subjected to severe stress and extensive killing may occur even with the best procedures available; for strains of *Paramecium* survival is usually a small fraction of 1%. Samples of many genera examined shortly after thawing may contain some motile cells, apparently intact nonmotile cells or ghosts, and debris. Many "intact" cells are nonviable. However, samples should not be discarded for several days because cells may recover motility as late as 3 or 4 days after thawing.

A flow chart depicting the steps used at the ATCC to select a procedure for cyropreserving particular protozoa is presented in Figure 2.

For explicit procedures for several dozen different protozoa and some algal and fungal protists, see Lee and Soldo (1992). Several genera have been recently cryopreserved, or successfully dried, at the ATCC but do not yet appear in their catalog. Other descriptions of methods for cyropreserving protozoa include those for *Naegleria* (Brown and Day, 1993), for *Babesi* vaccine (Hentrich and Bose, 1993), for *Entodinium* (Marcin *et al.*, 1992), for *Plasmodium* (Margos *et al.*, 1992), and for *Toxoplasma* (Popescu and Steriu, 1993). Public collections of protists are listed below.

Culture Repositories

Culture Collection Sources

American Type Culture Collection (ATCC)[1]
12301 Parklawn Dr.
Rockville, MD 20852-1776 USA
  (1000 protozoa, 1550 lower fungi)

Culture Collection for Algae and Protozoa (CCAP)[2]
Institute of Freshwater Ecology
The Windemere Laboratory
Far Sawrey, Ambleside
Cumbria LA22 OLP, U.K.
  (385 protozoa)

---

[1] At ATCC, ~130 additional protist strains should be placed in the algae according to Cavalier-Smith (1993). There are many public collections of fungi, some of which include slime molds, etc., that are often placed among the protozoa (see Table 2). From Nerad (1993).

[2] From European Culture Collections' Organization (1992). Information on Holdings and Services, ICECC, Publication No. 4.

Czechoslovak National Collection of Type Cultures (CNCTC)[2]
National Institute of Public Health
Srobarova 48
CS-100 42 Prague 10, Czechoslovakia
   (15 protozoa)

Collection Nationale de Cultures de Microorganismes (CNCM)[2]
Institut Pasteur
25, rue dw Docteur Roux
F-75724 Paris, Cedex 15, France
   (5 protozoa)

Computerized Networks

Microbial Information Network Europe (MINE)[2]
   (Separate node in each of 11 participating European countries; information on
   bacteria, fungi, and yeasts)

Microbial Strain Data Network (MSDN)[2]
307 Huntingdon Road
Cambridge CB3 OJX, U.K.
Internet: MSDN&CGNET.COM
   (Worldwide network; all types of protists and prokaryotes)

References

Adolph, K. W., ed. (1994). "Molecular Microbiology Techniques," Vol. 3, Part A. Plenum,
    New York.
Aitken, M. D., Heck, P. E., Mines, R. D., and Sherrard, J. H. (1992). *Water Environ. Res.*
    64(4), 347–358.
Alves, A. M. B., de Almeida, D. F., and von Krüger, W. M. A. (1994). Changes in *Trypsanosoma
    cruzi* kinetoplast DNA minicircles induced by environmental conditions and subcloning.
    J. Eukaryotic Microbiol. **41**(4), 415–419.
American Medical Television and American Society of Microbiology (1993). "The Polymerase
    Chain Reaction in Clinical and Laboratory Medicine." ASM Press, Washington, DC.
Anderson, O. R. (1988). "Comparative Protozoology: Ecology, Physiology, Life History."
    Springer-Verlag, New York.
Anderson, O. R. (1992). A fine structure study of *Physarum polycephalum* during transforma-
    tion from sclerotium to plasmodium. *J. Protozool.* **39**, 213–223.
Asmuth, D. M., Degirolami, P. C., Federman, M. *et al.* (1994). Clinical features of microsporidia
    in patients with AIDS. *Clin. Infect. Dis.* **18**(5), 819–825.
Avila, J. L., and Harris, J. R. (eds.) (1992). "Subcellular Biochemistry." Vol. 18: Intracellular
    Parasites. Plenum Press, NY.
Bard, D., and Lambrozo, J. (1992). Free-living amoebas causing meningoencephalitis and
    encephalitis: A review. *Med. Malad. Infect.* **22**(8–9), 698–705.
Basco, L. K., and Le Bras, J. (1994). *In vitro* activity of mitochondrial ATP synthetase inhibitors
    against *Plasmodium falciparum*. *J. Eukaryotic Microbiol.* **41**(3), 179–183.

Bates, P. A., Robertson, C. D., and Coombs, G. H. (1994). Expression of cysteine proteinases by metacyclic promastigotes of Leishmania mexicana. *J. Eukaryotic Microbiol.* **41**(3), 199–203.

Blackburn, E. H. (1986). Telomeres. In "The Molecular Biology of Ciliated Protozoa" (J. G. Gall, ed.), pp. 155–178. Academic Press, New York.

Bochstahler, K. R., and Coats, D. W. (1993). Spatial and temporal aspects of mixotrophy in Chesapeake Bay dinoflagellates. *J. Eukaryotic Microbiol.* **40**(1), 49–60.

Breteler, W. C. M. K., and Laan, M. (1993). An apparatus for automatic counting and controlling density of pelagic food particles in cultures of marine organisms. *Mar. Biol.* **116**(1), 169–174.

Breznak, J. A., and Brune, A. (1994). Role of microorganisms in the digestion of lignocellulose by termites. *Annu. Rev. Entomol.* **39**, 453–487.

Brown, S., and Day, J. G. (1993). An improved method for the long term preservation of *Naegleria gruberi. Cryo-Letters* **14**(6), 349–354.

Carrick, H. J., Fahnenstiel, G. L., and Taylor, W. D. (1992). Growth and production of planktonic protozoa in Lake Michigan: *In situ* versus *in vitro* comparisons and importance to food web dynamics. *Limnol. Oceanogr.* **37**(6), 1221–1235.

Cavalier-Smith, T. (1993). Kingdom Protozoa and its eighteen phyla. *Microbiol. Rev.* **57**(4), 953–994.

Cech, T. R. (1986). Ribosomal RNA gene expression in *Tetrahymena:* Transcription and RNA splicing. In "The Molecular Biology of Ciliated Protozoa" (J. G. Gall, ed.), pp. 203–225. Academic Press, New York.

Chiesa, R., Silva, W. I., and Renaud, F. L. (1993). Pharmacological characterization of an opioid receptor in the ciliate *Tetrahymena. J. Eukaryotic Microbiol.* **40**(6), 800–804.

Comley, J. C. W., and Sterling, A. M. (1994). Artificial infections of *Pneumocystis carinii* in the SCID mouse and their use in the in vivo evaluation of antipneumocystis drugs. *J. Eukaryotic Microbiol.* **41**(6), 540–546.

Coombs, G. H., and North, M. J., eds. (1991). "Biochemical Protozoology." Taylor & Francis, New York.

Cooperstock, M., Dupont, H. L., Corrado, M. L., Fekety, R., and Murray, D. M. (1992). Evaluation of new anti-infective drugs for the treatment of diarrhea caused by *Giardia lamblia. Clin. Infect. Dis.* **15**(1), 244–248.

Cordovilla, P., Valdivia, E., Gonzalez-Segura, A., *et al.* (1993). Antagonistic action of the bacterium *Bacillus licheniformis* M-4 toward the amoeba *Naegleria fowleri. J. Eukaryotic Microbiol.* **40**(3), 323–328.

Corliss, J. O. (1984). The kingdom Protista and its 45 phyla. *Bio Systems* **17,** 87–126.

Corliss, J. O. (1994). An interim utilitarian ("user friendly") hierarchical classification and characterization of the protists. *Acta Protozool.* **33**(1), 1–51.

Curley, G. P., O'Donovan, S. M., McNally, J. *et al.* (1994). Aminopeptidases from *Plasmodium falciparum, Plasmodium chabaudi chabaudi* and *Plasmodium berghei. J. Eukaryotic Microbiol.* **41**(2), 119–123.

Dalrymple, B. P. (1993). Molecular variation and diversity in candidate vaccine antigens from *Babesia. Acta Trop.* **53**(3–4), 227–238.

Das, P., Sengupta, K., Pal, S., Das, D., and Pal, S. C. (1993). Biochemical and immunological studies on soluble antigens of *Entamoeba histolytica. Parasitol. Res. (Berlin)* **79**(5), 365–371.

Diamond, L. S., Clark, C. G., and Cunnick, C. C. (1995). YI-S, a casein-free medium for axenic cultivation of *Entamoeba histolytica,* related *Entamoeba, Giardia intestinalis,* and *Trichomonas vaginalis. J. Eukaryotic Microbiol.* **42**(3), 277–278.

Dickerson, H. W., Clark, T. G., and Findly, R. C. (1989). *Ichthyophthirius multifiliis* has membrane-associated immobilization antigens. *J. Protozool.* **36**(2), 159–164.

Dini, F., and Nyberg, D. (1993). Sex in ciliates. *Adv. Microbiol. Ecol.* **13**, 85–153.

Dirie, M. F., Murphy, N. B., and Gardiner, P. R. (1993). DNA fingerprinting of *Trypanosoma vivax* isolates rapidly identifies intraspecific relationships. *J. Eukaryotic Microbiol.* **40**(2), 132–134.

Doerder, F. P., and Berkowitz, M. S. (1986). Purification and partial characterization of the H Immobilization antigens of *Tetrahymena thermophila. J. Protozool.* **33**(2), 204–208.

Erdmann, V. A., and Wolters, J. (1988). Collection of published 5S, 5.8S, and 4.5S ribosomal RNA sequences. *Nucleic Acids Res.* [Suppl.] **16**, r1–r70.

European Culture Collections Organization (1992). Information on Holdings and Services, ICECC, Publication No. 4.

Farthing, M. J. G., Varon, S. R., and Keusch, G. T. (1983). Mammalian bile promotes growth of *Giardia lamblia* in axenic culture. *Trans. R. Soc. Trop. Med. Hyg.* **77**, 467–469.

Fenchel, T. (1987). "Ecology of Protozoa: The Biology of Free-living Phagotrophic Protists." Sci. Tech. Publ., Madison, WI.

Fenchel, T. (1993). Methanogenesis in marine shallow water sediments: the quantitative role of anaerobic protozoa with endosymbiotic methanogenic bacteria. *Ophelia* **37**(1), 67–82.

Fernandez Galiano, D. (1994). The ammoniacal silver carbonate method as a general procedure in the study of protozoa from sewage and other waters. *Water Res.* **28**(2), 495–496.

Fernandez-Leborans, G., and Novillo, A. (1992). Hazard evaluation of lead effects using marine protozoan communities. *Aquat. Sci.* **54**(2), 128–140.

Fleury, A., and Laurent, M. (1994). Transmission of surface pattern through a dedifferentiated stage in the ciliate *Paraurostyla.* Evidence from the analysis of microtubule and basal body deployment. *J. Eukaryotic Microbiol.* **41**(3), 276–291.

Foissner, W. (1987). Soil protozoa: Fundamental problems, ecological significance, adaptations in ciliates and testaceans, bioindicators and guide to the literature. *Prog. Protistol.* **2**, 69–212.

Forge, T. A., Berrow, M. L., Darbyshire, J. F., and Warren, A. (1993). Protozoan bioassays of soil amended with sewage sludge and heavy metals, using the common soil ciliate *Colpoda steinii. Biol. Fertil. Soils* **16**(4), 282–286.

Frankel, J. (1989). "Pattern Formation: Ciliate Studies and Models." Oxford Univ. Press, New York.

Fulton, C. (1993). *Naegleria:* A research partner for cell and developmental biology. *J. Eukaryotic Microbiol.* **40**(4), 520–532.

Gall, J. G., ed. (1986). "The Molecular Biology of Ciliated Protozoa." Academic Press, New York.

Garcia, L. S., and Bruckner, D. A. (1993). "Diagnostic Medical Parasitology," 2nd ed. ASM Press, Washington, DC.

Gasol, J. M. (1993). Benthic flagellates and ciliates in fine freshwater sediments: Calibration of a live counting procedure and estimation of their abundances. *Microb. Ecol.* **25**(3), 247–262.

Gauthier, J. D., Feig, B., and Vasta, G. R. (1995). Effect of fetal bovine serum glycoproteins on the in vitro proliferation of the oyster parasite *Perkinsus marinus:* development of a fully defined medium. *J. Eukaryotic Microbiol.* **42**(3), 307–313.

Ghilardi, M., Christensen, S. T., Schousboe, P., and Rasmussen, L. (1992). Compounds stimulating growth and multiplication in ciliates: Do also free-living cells release growth factors? *Naturwissenschaften* **79**(5), 234–235.

Ginsburg, G. T., Gollop, R., Yu, Y. *et al.* (1995). The regulation of *Dictyostelium* development by transmembrane signalling. *J. Eukaryotic Microbiol.* **42**(3), 200–205.

Glick, B. R., and Pasternak, J. J. (1994). "Molecular Biotechnology: Principles and Applications." ASM Press, Washington, DC.

Gold, D. (1993). *Trichomonas vaginalis* strain differences in adhesion to plastic and virulence *in vitro* and *in vivo*. *Parasitol. Res.* (*Berlin*) **79**(4), 309–315.

Guerrini, F., Segur, C., Gargani, D., Tibayrene, M., and Dollet, M. (1992). An isozyme analysis of the genus *Phytomonas*: Genetic, taxonomic and epidemiologic significance. *J. Protozool.* **39** (4), 516–521.

Hadas, E., and Mazur, T. (1993). Biochemical markers of pathogenicity and virulence of *Acanthamoeba* sp. strains. *Parasitol. Res.* (*Berlin*) **79**(8), 696–698.

Haeckel, E. (1866). "Generelle Morphologie der Organismen." Reimer, Berlin.

Hague, R., Kress, K., Wood, S. *et al.* (1993). Diagnosis of pathogenic *Entamoeba histolytica* infection using a stool ELISA based on monoclonal antibodies to the galactose-specific adhesion. *J. Infect. Dis.* **167**(1), 247–249.

Hall, B. F., and Joiner, K. A. (1993). Developmentally regulated virulence factors of *Trypanosoma cruzi* and their relationship to evasion of host defenses. *J. Eukaryotic Microbiol.* **40**(2), 207–213.

Hall, J. A., Barrett, D. P., and James, M. R. (1993). The importance of phytoflagellate, heterotrophic flagellate and ciliate grazing on bacteria and picoplankton sized prey in a coastal marine environment. *J. Plankton Res.* **15**(9), 1075–1086.

Harris, E. H. (1989). "The Chlamydomonas Source Book." Academic Press, San Diego, CA.

Harvey, H. R., and McManus, G. B. (1991). Marine ciliates as a widespread source of tetrahymenol and hopan-3-beta-ol in sediments. *Geochim. Cosmochim. Acta* **55**(11), 3387–3390.

Hellung-Larsen, P., Lynne, I., Andersen, A. P., and Koppelhus, U. (1993). Characteristics of dividing and non-dividing *Tetrahymena* cells at different physiological states. *Eur. J. Protistol.* **29**(2), 182–190.

Hentrich, B., and Bose, R. (1993). Cryopreservation of *Babesia divergens* from jirds as a live vaccine for cattle. *Internatl. J. Parasitol.* **23**(6), 771–776.

Hirumi, H., Hirumi, K., Moloo, S. K., and Shaw, M. K. (1992). *Trypanosoma brucei brucei: In vitro* production of metacyclic forms. *J. Protozool.* **39**(5), 619–627.

Holler, S., Galle, A., and Pfennig, M. (1994). Degradation of food compounds and growth response on different food quality by the anaerobic ciliate *Trimyema compressum. Arch. Microbiol.* **161**(1), 94–98.

Horstmann, R. D., Leippe, M., and Tannick, E. (1992). Host tissue destruction by *Entamoeba histolytica*—molecules mediating adhesion, cytolysis and proteolysis. Mem. Inst. Oswaldo Cruz 87 (S5):57–60.

Hyde, J. E. (ed.) (1993). *Protocols in Molecular Parasitology.* Vol. 21, *Methods in Molecular Biology* Series. 480 pp. Hurana Press, Totowa, NJ.

Iltzsch, M. H. (1993). Pyrimidine salvage pathways in *Toxoplasma gondii. J. Eukaryotic Microbiol.* **40**(1), 24–28.

Ishii, A., Matsuoka, H., Aji, T. *et al.* (1994). Parasite infection and cancer—with special emphasis on *Schistosoma japonicum* infection—a review. *Mutat. Res.* **305**(2), 273–281.

Jankevicius, S. I., de Almeida, M. L., Jankevicius, I. V. *et al.* (1993). Axenic cultivation of trypanosomatids found in corn (*Zea mays*) and in phytophagus hermipterans (*Leptaglossus zonatus* Coreidae) and their experimental transmission. *J. Eukaryotic Microbiol.* **40**(5), 576–581.

Jerome, C. A., Montagnes, D. J. S., and Taylor, F. J. R. (1993). The effect of quantitative protargol stain and Lugol's and Bouin's fixatives on cell size—a more accurate estimate of ciliate species biomass. *J. Eukaryotic Microbiol.* **40**(3), 254–259.

John, D. T., and Howard, M. J. (1993). Virulence of *Naegleria fowleri* affected by axenic cultivation and passage in mice. *Folia Parasitol.* **40**(3), 187–191.

Joiner, K. A., and Dubremetz, J. F. (1993). *Toxoplasma gondii:* A protozoan for the nineties. *Infect. Immun.* **61**(4), 1169–1172.

Kaidoh, T., Nath, J., Fujioka, H., Okoye, V., and Aikawa, M. (1995). Effect and localization of trifluralin in *Plasmodium falciparum* gametocytes: An electron microscopic study. *J. Eukaryotic Microbiol.* **42**(1), 61–64.

Kalinina, L. V., and Page, F. C. (1992). Culture and preservation of naked amoebae. *Acta Protozool.* **31**(2), 115–126.

Kaneshino, E. S., Wyder, M. A., Zhou, L. H. *et al.* (1993). Characterization of *Pneumocystis carinii* preparations developed for lipid analysis. *J. Eukaryotic Microbiol.* **40**(6), 805–815.

Kessler, D. R., and Levine, H. (1993). Pattern formation in *Dictyostelium* via the dynamics of cooperative biological entities. *Physiol. Rev. E* **48**(6), 4801–4804.

Kierszenbaum, F., ed. (1994). "Parasitic Infections and the Immune System." Plenum, New York.

Klein, R., and Groliere, C. A. (1993). Enantiomeric separation of biopterin and neopterin stereoisomers by chiral HPLC-application to the identification of natural pterins in protozoa. *Chromatographia* **36**, 71–75.

Komala, Z. (1993). Response of *Paramecium primaurelia* on the air pollution. *Folia Biol. (Krakow)* **41**(1–2), 17–23.

Kreier, J. P., ed. (1991–1994). "Parasitic Protozoa," Vols. 1–4, Vols. 6–8. Academic Press, San Diego, CA.

Kung, C., Preston, R. R., Maley, M. E. *et al.* (1992). *In vivo Paramecium* mutants show that calmodulin orchestrates membrane responses to stimuli. *Cell Calcium* **13**(6–7), 413–425.

Laybourn-Parry, J. (1992). "Protozoan Plankton Ecology." Chapman & Hall, New York.

Lee, C. H., Bauer, N. L., Shaw, M. M. *et al.* (1993). Proliferation of rat *Pneumocystis carinii* on cells sheeted on microcarrier beads in spinner flasks. *J. Clin. Microbiol.* **31**(6), 1659–1662.

Lee, J. J. (1985a). Trichomonadida. *In* "An Illustrated Guide to the Protozoa" (J. J. Lee, S. H. Hutner, and E. C. Bovee, eds.), pp. 119–127. Society of Protozoologists, Lawrence, KS.

Lee, J. J. (1985b). Diplomonadida. *In* "An Illustrated Guide to the Protozoa" (J. J. Lee, S. H. Hutner, and E. C. Bovee, eds.), pp. 130–134. Society of Protozoologists, Lawrence, KS.

Lee, J. J., and Hutner, S. H. (1985). Kinetoplastida. *In* "An Illustrated Guide to the Protozoa" (J. J. Lee, S. H. Hutner, and E. C. Bovee, eds.), pp. 141–155. Society of Protozoologists, Lawrence, KS.

Lee, J. J., Hutner, S. H., and Bovee, E. C., eds. (1985). "An Illustrated Guide to the Protozoa." Society of Protozoologists, Lawrence, KS.

Lee, J. J., and Soldo, A. T., eds. (1992). "Protocols in Protozoology." 148 sections. Society of Protozoologists, Allen Press, Lawrence, KS.

Lee, R. E., and Kugrens, P. (1992). Relationship between the flagellates and the ciliates. *Microbiol. Rev.* **56**(4), 529–542.

Levandowsky, M., and Hutner, S. H., eds. (1979–1981). "Biochemistry and Physiology of the Protozoa," 2nd ed., Vols. 1–4. Academic Press, New York.

Levine, N. D. (1985). Phylum II. Apicomplexa, Levine, 1970. *In* "An Illustrated Guide to the Protozoa" (J. J. Lee, S. H. Hutner, and E. C. Bovee, eds.), pp. 322–374. Society of Protozoologists, Lawrence, KS.

Levine, N. D., Corliss, J. O., Cox, F. E. G. *et al.* (1980). A newly revised classification of the protozoa. *J. Protozool.* **27**, 37–59.

Lim, E. L., Amaral, L. A., Caron, D. A., and Delong, E. F. (1993). Application of rRNA based probes for observing marine nanoplanktonic protists. *Appl. Environ. Microbiol.* **59**(5), 1647–1655.

Lotter, H., Mannweiler, E., Schreiber, M., and Tannich, E. (1992). Sensitive and specific serodiagnosis of invasive amoebiasis by using a recombinant surface protein of pathogenic *Entamoeba histolytica. J. Clin. Microbiol.* **30**(12), 3163–3167.

Lovejoy, C., Vincent, W. F., Frenette, J. S., and Dodson, J. J. (1993). Microbial gradients in a turbid estuary—Application of a new method for protozoan community analysis. *Limnol. Oceanogr.* **38**(6), 1295–1303.

Lujan, H. D., and Nash, T. E. (1994). The uptake and metabolism of cysteine by *Giardia lamblia* trophozoites. *J. Eukaryotic Microbiol.* **41**(2), 169–175.

Luporini, P., Vallesi, A., Miceli, C., and Bradshaw, R. A. (1995). Chemical signaling in ciliates. *J. Eukaryotic Microbiol.* **42**(3), 208–212.

Machemer, H. (1987). From structure to behavior: *Stylonichia* as a model system for cellular behavior. *Prog. Protistol.* **2**, 213–330.

Madoni, P. (1994). A sludge biotic index (SBI) for the evaluation of the biological performance of activated sludge plants based on microfauna analysis. *Water Res.* **28**(1), 67–75.

Maizels, R. M., Blaxter, M. L., Robertson, B. D., and Selkirk, M. E. (1992). "Parasite Antigens, Parasite Genes: A Laboratory Manual for Molecular Parasitology" Press Syndicate, University of Cambridge, England.

Marche, S., Roth, C., Manohar, S. K., Dollet, M., and Baltz, T. (1993). RNA virus-like particles in pathogenic plant trypanosomatids. *Mol. Biochem. Parasitol.* **57**(2), 261–268.

Marciano-Cabral, F., and Toney, D. M. (1994). Modulation of biological functions of *Naegleria fowleri* amoebae by growth medium. *J. Eukaryotic Microbiol.* **41**(1), 38–46.

Marcin, R., Kisidayova, S., Fejes, J., Zelenak, I., and Kmet, V. (1992). A simple technique for cryopreservation of the rumen protozoan *Entodinium caudatiium. Cryo-Letters* **13**(3), 175–182.

Margos, G., Maier, W. A., and Seitz, H. M. (1992). Experiments on cryopreservation of *Plasmodium falciparum. Trop. Med. Parasitol.* **43**(1), 13–16.

Margulis, L. (1974). Five kingdom classification and the origin and evolution of cells. *Evol. Biol.* **7**, 45–48.

Mehlotra, R. K., and Shukla, O. P. (1993). *In vitro* susceptibility of *Acanthamoeba culburtsoni* to inhibitors of folate biosynthesis. *J. Eukaryotic Microbiol.* **40**(1), 14–17.

Meyer, E. B., and Nanney, D. L. (1987). Isozymes in the ciliated protozoan *Tetrahymena. Curr. Top. Biol. Med. Res.* **13**, 61–101.

Meyer, F., Schmidt, H. J., Plümper, E. *et al.* (1991). UGA is translated cysteine in pheromone 3 of *Euplotes octocarinatus. Proc. Natl. Acad. Sci. USA* **88**, 3758–3761.

Michalowski, T., Muszynski, P., and Landa, I. (1991). Factors influencing the growth of rumen ciliates, *Eudiplodium maggii in vitro. Acta Protozool.* **30**(2), 115–120.

Moulin, A. M., Darcy, F., Sibley, L. D. *et al.* (1993). The immunobiology of toxoplasmosis—discussion. *Res. Immunol.* **144**(1), 68–79.

Mukherjee, R. M., Bhol, K. C., Mehra, S., Maitra, T. K., and Jalan, K. N. (1993). Zymodeme alteration of *Entamoeba histolytica* isolates under varying conditions. *Tr. Roy. Soc. Trop. Med. Hyg.* **87**(4), 490–491.

Nanney, D. L., Parks, C., and Simon, E. M. (1995). In preparation.

Nerad, T. A., ed. (1993). "ATCC Catalogue of Protists," 18th ed. American Type Culture Collection, Rockville, MD.

Nisbet, B. (1984). "Nutrition and Feeding Strategies in Protozoa." Croom-Helm, London.

Noever, D. A., Matsos, H. C., and Looger, L. L. (1992). Bioconvective indicators in *Tetrahymena:* Nickel and copper protection from cadmium poisoning. *Environ. Sci. Eng.* **27**(2), 403–417.

Novotny, J., and Weiser, J. (1993). Transovarial transmission of *Nosema lymantriae* (protozoa, Microsporidia) in the gypsy moth *Lymantria dispar.* L. *Biologia (Bratislava)* **48**(2), 125–129.

Ohman, M. D. (1992). Immunological recognition of oligotrich ciliates. *Mar. Biol.* **114**(4), 653–660.

Orias, E., and Bruns, P. J. (1976). Induction and isolation of mutants in *Tetrahymena. Methods Cell Biol.* **13**, 247–282.

Osuna, A., Rodriguez-Cabezas, N., Gamarro, F., and Mascaro, C. (1994). The different behavior in diphtheria toxin, modeccin and ricin in Hela cells infected with *Trypanosoma cruzi. J. Eukaryotic Microbiol.* **41**(3), 231–236.

Pan, A. A., Duboise, S. M., Eperon, S. *et al.* (1993). Developmental life cycle of *Leishmania*—Cultivation and characterization of cultured extracellular amastigotes. *J. Eukaryotic Microbiol.* **40**(2), 213–223.

Pidherney, M. S., Olizadeh, H., Stewart, G. L., McCulley, J. P., and Niederkorn, J. Y. (1993). *In vitro* and *in vivo* tumoricidal properties of a pathogenic free-living amoeba. Cancer Letters **72**(1–2), 91–98.

Popescu, C., and Steriu, D. (1993). Long term cryopreservation of *Toxoplasma gondii. Cryo-Letters* **14**(6), 383–388.

Preer, J. R., Jr., Preer, L. B., Rudman, B., and Barnett, A. (1987). Molecular biology of the genes for immobilization antigens in *Paramecium. J. Protozool.* **34**(4), 418–423.

Presber, W. (1994). New drugs against malaria parasites. *Immun. Infekt.* **22**(2), 48–52.

Prescott, D. M. (1992). The unusual organization and processing of genomic DNA in hypotrichous ciliates. *Trends Genet.* **8**(12), 439–445.

Qureshi, M. N., Perez, A. A., Madayag, R. M., and Bottone, E. J. (1993). Inhibition of *Acanthamoeba* spp. by *Pseudomonas aeruginosa*—rationale for their selective exclusion in corneal ulcers and contact lens care systems. *J. Clin. Microbiol.* **31**(7), 1908–1910.

Raper, K. B. (1984). "The Dictyostelids." Princeton Univ. Press, Princeton, NJ.

Reduth, D., Grootenhuis, J. G., Olubayo, R. O. *et al.* (1994). African buffalo serum contains novel trypanocidal protein. *J. Eukaryotic Microbiol.* **41**(2), 95–103.

Renaud, F. L., Colon, I., Lebron, J. *et al.* (1995). A novel opioid mechanism seems to modulate phagocytosis in *Tetrahymena. J. Eukaryotic Microbiol.* **42**(3), 205–207.

Rojas, M. O., and Wasserman, M. (1993). Effect of low temperature on the *in vitro* growth of *Plasmodium falciparum. J. Eukaryotic Microbiol.* **40**(2), 149–152.

Rubin, H. (1994). Epigenetic nature of neoplastic transformation. *In* "Developmental Biology and Cancer" (G. M. Hodges and C. Rowlatt, eds.), pp. 61–84. CRC Press, Boca Raton, FL.

Saffer, L. D., Mercereau-Puijalon, O., Dubremetz, J. F., and Schwartzman, J. D. (1992). Localization of a *Toxoplasma gondii* rhoptry protein by immunoelectron microscopy during and after host cell penetration. *J. Protozool.* **39**(4), 526–530.

Sam-Yellowe, T. Y. (1992). Molecular factors responsible for host cell recognition and invasion in *Plasmodium falciparum. J. Protozool.* **39**(1), 181–189.

Sam-Yellowe, T. Y., Fujioka, H., Aikawa, M., and Messineo, D. G. (1995). *Plasmodium falciparum* rhoptry proteins of 140/130/110 kd (Rhop-H) are located in an electron lucent compartment in the neck of the rhoptries. *J. Eukaryotic Microbiol.* **42**(3), 224–231.

Sanchez-Moreno, M., Fernandez-Becerra, C., Entrala, E. *et al.* (1995). *In vitro* culture of *Phytomonas sp.* isolated from *Euphoria characias*. Metabolic studies by ^{1}H NMR. *J. Eukaryotic Microbiol.* **42**(3), 314–320.

Sekkat, N., Ledu, A., Jounay, J. M., and Guerbet, M. (1992). Study of the interactions between copper, cadmium, and Ferbam using the protozoan *Colpidium campylum* bioassay. *Ecotoxicol. Environ. Saf.* **24**(3), 294–300.

Shakibaei, M., and Frevent, U. (1992). Cell surface interactions between *Trypanosoma congolense* and macrophages during phagocytosis *in vitro. J. Protozool.* **39**(1), 224–235.

Sibley, L. D., Pfefferkorn, E. R., and Boothroyd, J. C. (1993). Development of genetic systems for *Toxoplasma gondi. Parasitol. Today* **9**(10), 392–395.

Simione, F. P., and Brown, E. M., eds. (1991). "ATCC Preservation Methods: Freezing and Freeze-drying," 2nd ed. American Type Culture Collection, Rockville, MD.

Smith, D. L., Berkowitz, M. S., Potoczak, D. *et al.* (1992). Characterization of the T, L, I, S, M and P cell surface (immobilization) antigens of *Tetrahymena thermophila:* Molecular weights, isoforms and cross-reactivity of sera. *J. Protozool.* **39**(3), 420–428.

Society of Protozoologists. (1994). Third International Workshops on Opportunistic Protists. Cleveland State University, Cleveland, OH, June 24–29, 1994. *J. Eukaryotic Microbiol.* **41**(5), 1S–128S.

Sonneborn, T. M. (1939). *Paramecium aurelia;* Mating types and groups, lethal interactions; determination and inheritance. *Am. Nat.* **73**, 390–413.

Sonneborn, T. M. (1950). Methods in the general biology and genetics of *Paramecium aurelia.* *J. Exp. Zool.* **113**, 87–148.

Souto-Padron, T., Almeida, J. D., de Souza, W., and Travassos, L. R. (1994). Distribution of α-galactosyl-containing epitopes on *Trypanosoma cruzi* trypomastigote and amastigote forms from infected Vera cells detected by Chagasic antibodies. *J. Eukaryotic Microbiol.* **41**(1), 47–54.

Sparagano, O. (1993). Differentiation of *Naegleria fowleri* and other naeglerias by polymerase chain reaction and hybridization methods. *FEMS Microbiol. Lett.* **110**(3), 325–330.

Srikanth, S., and Berk, S. G. (1993). Stimulatory effect of cooling tower biocides on amoebae. *Appl. Environ. Microbiol.* **59**(10), 3245–3249.

Stallwitz, E., and Hader, D. P. (1993). Motility and phototactic orientation of the flagellate *Euglena gracilis* impaired by heavy metal ions. *J. Photochem. Photobiol. B* **18**(1), 67–74.

Starink, M., Bargilisen, M. J., Bak, R. P. M., and Cappenberg, T. E. (1994). Quantitative centrifugation to extract benthic protozoa from freshwater sediments. *Appl. Environ. Microbiol.* **60**(1), 167–173.

Steele, C. J., Barkocygallagher, G. A., Preer, L. B., and Preer, J. R. (1994). Developmentally excised sequences in micronuclear DNA of *Paramecium. Proc. Natl. Acad. Sci. U.S.A.* **91**(6), 2255–2259.

Stringer, J. R., Stringer, S. L., Zhang, J. *et al.* (1993). Molecular genetic distinction of *Pneumocystis carinii* from rats and humans. *J. Eukaryotic Microbiol.* **40**(6), 733–741.

Sugiura, K. (1992). A multispecific laboratory microcosm for screening ecotoxicological impacts of chemicals. *Environ. Toxicol. Chem.* **11**(9), 1217–1226.

Teixeira, M. M. G., Campaner, M., and Camargo, E. P. (1995). Characterization of the target antigens of *Phytomonas* specific monoclonal antibodies. *J. Eukaryotic Microbiol.* **42**(3), 232–237.

Thomford, J. W., Conrad, P. A., Boyce, W. M., Holman, P. J., and Jessup, D. A. (1993). Isolation and *in vitro* cultivation of *Babesia* parasites from free-ranging desert bighorn sheep (*Ovis canadensis nelsoni*) and mule deer (*Odocoileus hemionus*) in California. *J. Parasitol.* **79**(1), 77–84.

Toney, D. M., and Marciano-Cabral, F. (1994). Modulation of complement resistance and virulence of *Naegleria fowleri* amoebae by alterations in growth media. *J. Eukaryotic Microbiol.* **41**(4), 337–343.

Toews, S., Beverleyburton, M., and Lawrimore, T. (1993). Helminth and protist parasites of zebra mussels, *Dreissena polymorpha* (Pallas, 1771) in the Great Lakes region of southwestern Ontario, with comments on associated bacteria. *Canad. J. Zool.* **71**(9), 1763–1766.

Turner, M. J., and Arnot, D., eds. (1988). "Molecular Genetics of Parasitic Protozoa." Cold Spring Harbor Lab., Cold Spring Harbor, NY.

Uliana, S. R. B., Nelson, K., Beverley, S. M., Camargo, E. P., and Floeter-Winter, L. M. (1994). Discrimination amongst Leishmania by polymerase chain reaction and hybridization with small subunit ribosomal RNA derived oligonucleotides. *J. Eukaryotic Microbiol.* **41**(4), 324–330.

Valentin, A., Precigout, E., and Lhostis, H. (1993). Cellular and humoral immune responses induced in cattle by vaccination with *Babesia divergens* culture-derived exoantigens correlate with protection. *Infect. Immun.* **61**(2), 734–741.

Van Eys, G. J. J. M., Schoone, G. J., Kroon, N. C. M., and Ebeling, S. B. (1992). Sequence analysis of small subunit ribosomal RNA genes and its use for detection and identification of *Leishmania parasites. Mol. Biochem. Parasitol.* **51**(1), 133–142.

Van Houten, J. (1994). Chemosensory transduction in eukaryotic microorganisms. *Trends Neurosci.* **17**(2), 62–71.

Vickerman, K. (1992). The diversity and ecological significance of protozoa. *Biodiversity Conserv.* **1**(4), 334–341.

Visvesvara, G. S. (1993). Epidemiology of infections with free-living amoebas and laboratory diagnosis of microsporidiosis. *Mt. Sinai J. Med.* **60**(4), 283–288.

Visvesvara, G. S., and Balamuth, W. (1975). Comparative studies on related free-living and pathogenic amoebae with special reference to *Acanthamoeba. J. Protozool.* **22**, 245–256.

Walzer, P. D., ed. (1993). "*Pneumocystis carinii* Pneumonia," 2nd ed. Dekker, New York.

Weiss, L. M., Laplace D., Takvorian, P. M., *et al.* (1995). A cell culture system for study of the development of *Toxoplasma gondii* bradyzoites. *J. Eukaryotic Microbiol.* **42**(2), 150–157.

Wheatley, D. H., Christensen, S. T., Schousboe, P., and Rasmussen, L. (1993). Signalling in cell growth and death: Adequate nutrition alone may not be sufficient for ciliates: A minireview. *Cell Biol. Int.* **17**(9), 817–823.

Wheeler-Alm, E., and Shapiro, S. Z. (1992). Evidence of tyrosine kinase activity in the protozoan parasite *Trypanosoma brucei. J. Protozool.* **39**(3), 413–416.

Whittaker, R. H. (1969). New concepts of kingdoms of organisms. *Science* **163**, 150–160.

Yang, S., Bergman, L. W., Scholl, D. R., and Rowland, E. C. (1994). Cloning of a partial length cDNA encoding the C-terminal portion of the 75-77 kDa antigen of *Trypanosoma cruzi. J. Eukaryotic Microbiol.* **41**(5), 435–441.

Yanagi, A. (1992). Behavior and DNA synthesis of the nuclei produced after meiosis in *Paramecium caudatum. Eur. J. Protistol.* **28**(1), 37–42.

Yasunaga, C., Funakoshi, M., and Kawarabata, T. (1994). Effects of host cell density on cell infection level in *Antheraea eucalypti* (Lepidoptera: Saturniidae) cell cultures persistently infected with *Nosema bombycis* (Microsporidia: Nosematidae). *J. Eukaryotic Microbiol.* **41**(2), 133–137.

Yasunaga, C., Inoue, S., Funakoshi, M., Kawarabata, T., and Hayasaka, S. (1995). A new method for inoculation of poor germinator, *Nosema sp.* N1S M11 (Microsporidia, Nosematidae) into an insect cell culture. *J. Eukaryotic Microbiol.* **42**(2), 191–195.

CHAPTER 8

Animal Cells in Culture

Robert J. Hay

Background

Introduction

Most of the other chapters in this volume deal with algae and microorganisms, whereas this chapter includes details relating generally to the diversity, utility, development, cryopreservation, and characterization of metazoan cells in culture. Interest in this technology has increased dramatically in the past 15–20 years as indicated by the publication of new and expanded textbooks (Adams, 1980; Butler, 1987; Freshney, 1987), methodological treatises such as *The Journal of Tissue Culture Methods and Cytotechnology,* review articles (Hink, 1979, 1989; Lydersen, 1987; Ho and Wang, 1991), and specialized periodicals such as *In Vitro Cell and Developmental Biology, Advances in Cell Culture, Animal Cell Biotechnology,* and *Cell and Tissue Research.* The distribution of cell lines from national repositories has also increased, reportedly more than 10-fold during this past interim (Hay, 1991).

Classification and Diversity

Current technology enables one to isolate, maintain, and/or cultivate virtually any tissue cells from any species of interest. Decisions are required at the outset to determine for the studies contemplated optimal species and tissue types, degree of homogeneity needed, longevity anticipated, plus

TABLE 1
Choice of Source Species for Cell Line Isolation and Establishment

Example species	Advantages	Disadvantages
Human	"Pertinence"; funding	Availability; limited longevity for normal cells
Rodent	Ready availability; inexpensive; manipulate source	Transform or exhibit karyotype change; endogenous viruses
Avian	Developmental stage availability Inexpensive Manipulate source tissue	Complicated cytogenetics Endogenous viruses Limited longevity for normal cells
Bovine	Size; inexpensive; ready availability	Karyotype change; bovine viral diarrhea virus infection

the structural and functional characteristics to be maintained. Choice of source species may be problematical, especially with regard to endogenous viruses. Other criteria for consideration include relevancy to human biology, karyotype stability, size of the organs, and so forth. Table 1 gives example choices with advantages and disadvantages.

Primary or secondary cultures might be selected if specific and necessary differentiated functions are required but are not available with continuous cell lines. Mammalian neuronal cell types can be cited as examples. Methods for isolation and maintenance of cells from the mammalian brain and peripheral nervous system are reasonably well established (Steinsvag, 1991; Cunha and Vitkovic, 1991).

An increasing array of normal cell types of limited doubling potential can now be offered. For example, human keratinocytes and corneal and prostatic epithelia can be cryopreserved in quantity, reconstituted, and propagated for many passages. Human stromal and endothelial cell lines are similarly available from the American Type Culture Collection (ATCC), Certified Cell Line (CCL), and Certified Repository Line (CRL) banks (e.g., ATCC CCL 75 and CRL 1730, respectively). Continuous lines from the desired species and tissue that express the function of interest are

TABLE 2
Representative Cell Lines[a]

ATCC no.	Designation	Example utility
CCL 75	WI-38	Vaccines; aging studies
CCL 171	MRC-5	Vaccines; diagnostics
CCL 212	MRC-9	Aging; virology
CCL 10	BHK-21	Transformation studies; virology
CCL 34	MDCK	Transport studies
CCL81	VERO	Virology; vaccines
CCL 61	CHO-K1	Genetics; engineering
CRL 1581	Sp2/0-Ag14	Nonsecreting fusion myeloma
CRL 1711	Sf9	Infect with Baculovirus expression vectors
CRL 1730	HUV-EC-C	Endothelium; function
CCL 226	$C_3H/10T1/2$	Carcinogen tests

[a] For more detail on these and other cell lines, see Hay *et al.* (1992).

frequently considered ideal. Examples include the Madin–Darby Canine Kidney line (MDCK ATCC CCL 34) and the 4MBr-5 monkey bronchial epithelial line (ATCC CCL 208). The cell line Sf9 (ATCC CRL 1711) from the ovary of a Fall armyworm, *Spodoptera frugiperda,* is especially useful for production of eukaryotic gene products (Fraser, 1989) due to the ease of handling and the relatively efficient levels of expression of inserted genes that can be effected in laboratory-scale production (averaging 1–10 μg per 10^6 cells). Advantageous posttranslational processing may occur in this eukaryotic system as opposed to engineered prokaryotes. However, some of these are remarkably different in insect cells as opposed to those from higher animals such as the Chinese hamster. Table 2 lists these and various other example lines to illustrate diversity. Genetically engineered lines are included in Table 3. Note that hybridomas are appropriately included in this general category, as are transduced and transfected lines plus cell lines for virus packaging.

Classification and nomenclature for cell lines are neither established nor systematic, as is the case for microorganisms discussed in other chapters of this volume. The Terminology Committee of The American Tissue Culture Association has considered the subject. General recommendations have been presented (Schaeffer, 1984), especially that no changes be made

TABLE 3
Example Genetically Engineered Cell Lines[a]

ATCC no.	Designation	Engineered property
CRL 1650	COS-1	Host line for recombinant SV40 propagation
CRL 8002[b]	OKT4	Produces monoclonal to CD4
CRL 8017[b]	H25B10	Produces monoclonal to hepatitis B virus
CRL 8179[b]	Cl.Ly1	Produces IL-3 and other cytokines
CRL 9078[b]	PA317	Cotransfected with pPAM3 and HSV TK- for retroviral packaging
CRL 9096[b]	CHO/dhFr	For selection after fusion
CRL 9606[b]	CHO 1.15$_{500}$	Produces human tissue plasminogen activator

[a] See Hay et al. (1992) for additional lines and information.
[b] Deposited for patent purposes.

to the cell line designation given by the originator without publication and rationale. Suggestions have been made that four-letter acronyms could be used to designate specific lines, or an initial two letters to indicate institution of origin, followed by numbers to specify identity. Examples include NCTC clone 929 for the National Cancer Institute Tissue Culture unit clone 929 of the L cell line and WI-38, Wistar Institute number 38. This system has not yet been adopted in any uniform manner, and many alternatives are in common use today.

Industrial Importance

For decades animal cells in culture have been exploited in many areas of research and for vaccine manufacture. Advances in somatic cell hybridization, genetic manipulation, transduction, and transfection now permit generation of new cell lines having tremendous potential utility, not only in the areas recognized formerly, but also for the production of monoclonal antibodies, biopharmaceuticals, and an array of cytokines (Sasaki and Ikura, 1991; Ho and Wang, 1991). Lines genetically engineered to produce such reagents can be expanded and propagated to bioreactors for large-scale production work. For example, substances such as tissue plasminogen activator, human insulin and growth hormone, α-interferon, OKT3 monoclonal antibody, erythropoietin, and recombinant hepatitis virus vaccine have all been approved by regulatory agencies for pharmaceutical use by the public. The release of very many more can be anticipated in the near future (Spier, 1991).

Cell lines of commercial utility are often deposited in impartial national repositories at the time of initiation of a patent application. Representative patented cell lines currently available are included in Table 3.

Methods for Preservation

General detail on methods and reference to more extensive descriptions will be included in the following discussions.

Isolation and Development of Cell Lines

One may isolate cells for culture by planting a small fragment with a suitable substrate and medium to allow cell outgrowth (explant culture). Alternatively, the tissue may be dissociated to yield component cells for cultivation. Specific cell types may be recovered initially by careful dissection of the tissue and/or by fractionation of the primary cell suspension. The first cultivated population is referred to as the primary culture (Figure 1,I).

For normal fixed, postmitotic cells such as neurons, heart cells, and multinucleated, skeletal muscle cells, maintenance *in vitro* is possible but propagation is not. Many other cell types (e.g., fibroblasts, epithelia, lymphoblasts) can be passed to yield secondary, tertiary, and even higher order

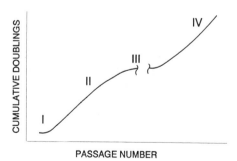

Figure 1. The evolution of a cell line is presented: I, the primary culture phase; II, the phase of rapid reproduction; III, the so-called crisis or senescent phase. In some cases (see text) the line can escape this phase to yield a continuous cell line (IV).

cultures, increasing the numbers of cells available at each passage (Figure 1,II). Some populations (e.g., most lymphoid and myelogenous cells) multiply as single cells or aggregates loosely attached or in suspension. Others, such as fibroblasts, adhere to the substrate to form a "monolayer," which requires dissociation at the end of each passage. Crude preparations of proteolytic enzymes such as trypsin or collagenase are commonly used for this purpose.

For normal cells as well as cells from some tumors, the rate of proliferation declines and the population may cease dividing or degenerate ("crisis," or senescence phase; Figure 1,III). The extent of proliferation varies with the source species and the age of the donor. Although the causes of this phenomenon are not understood, they are of considerable importance in cell and cancer biology as well as gerontology (Freshney, 1987).

In some cases, notably with murine and other rodent source tissues and tumors from various species, changes occurring during the crisis phase result in the appearance of a rapidly dividing population, a continuous cell line (Figure 1,IV). This may become tumorigenic after further passage or it may remain nontumorigenic, depending on culture conditions. The karyotype becomes grossly abnormal for murine cells but more subtle alterations may occur with cells from other species. Oncogenes or oncogenic viruses play a major role in the evolution of these lines.

Interestingly, the use of a serum-free system with supplemental insulin, transferrin, epidermal growth factor, high-density lipoprotein, and fibronectin enables apparently normal, murine, embryonic cells to escape the degenerative events and to form continuous cell lines (Loo *et al.*, 1987). This occurs without the more characteristic drop in proliferation and karyotype change. These authors thus propose that the inhibition of growth by serum factors and the selection of cells unresponsive to these factors contribute to the processes of crisis and immortalization as described.

Fibroblastic Cell Lines

Fibroblastic lines from skin, bronchus, lung, and colon of human and other animal origins (Table 4) have been produced in the author's laboratory using a standard explant culture method (Hay *et al.*, 1982). Tissue biopsies are collected, placed in a sterile nutrient medium containing antibiotics, and stored at 5°C until cultures could be prepared. Approximately 2–3 ml of (F12K or other formulas) serum-free medium is added to sterile T-flasks. After rinsing the surface of the flask floor, the medium is removed and the flasks are stoppered and set aside for use later. The biopsy (3–5 mm in diameter) is placed on the flat surface of an inverted 100-mm petri dish and rinsed with chemically defined medium containing antibiotics.

TABLE 4

Representaive Cell Lines Developed at the American Type
Culture Collection[a]

ATCC no.	Designation	Source/comment
Fibroblasts		
CCL 202	CCD-11Lu	Normal human lung/control
CRL 1475	CCD-27Sk	Normal human skin/control
CRL 1541	CCD-112CoN	Normal human colon/control
CRL 1139	EnSon	Skin/cystic fibrosis
CRL 1148	MaSan	Skin/Ehlers–Danlos syndrome
CRL 1485	CCD-32Lu	Lung/congenital heart disease
CRL 1498	CCD-39Lu	Lung/hyaline membrane disease
CCL 192	RFL6	Rat germ-free lung
CCL 195	AHL1	Armenian hamster lung
Epithelia		
CCL 208	4MBr-5	Normal rhesus bronchus
CRL 1790	CCD841CoN	Human colon/lacks keratin
CRL 1492	AR42J	Rat exocrine pancreatic tumor/functional
Hybridomas		
HB 155	Ep-16	Monoclonal to human keratinocyte surface antigen
HB172	10B9	Monoclonal to human endothelia and macrophages

[a] See Hay *et al.* (1992) or review the cell line database available through CODATA TYMNET for further information.

Tissue is minced into small fragments (usually 10 to 20) by means of sterile surgical blades, and the fragments are transferred to the previously wetted T-flasks using small-bore Pasteur pipettes. Usually 6 to 8 fragments are placed adjacent to each other in the center of the flask. The flask is inverted and 0.5 ml of medium, supplemented with 10% fetal bovine serum and antibiotics, is added. Each container is stoppered, inverted, and allowed to sit at room temperature for 30 min to assure good attachment of the fragments to the floor of the flask (initially in the upper position). After this, the flask is carefully turned back to its usual position to allow the medium to flow over the fragments on the flask floor without dislodging them and is placed at 37°C.

After 7 days of incubation the flasks are examined and the medium is renewed. After 14 days in culture the use of antibiotics is discontinued. The frequency of fluid renewal is determined by the amount of proliferation observed. Likewise the volume of medium used is progressively increased from the original 0.5 ml/T-flask to 3–4 ml/T-flask.

When satisfactory growth is evident, the fragments are removed from the flask and subcultures are prepared using a standard trypsin versene solution. Subsequent subcultures are made in larger flasks, which in turn are used to establish cultures in large (840 cm^2) roller bottles. The cells are harvested from these vessels, frozen, and stored in liquid nitrogen.

The lines so developed are useful in a very wide variety of metabolic studies, including work relating to matrix synthesis, genetic and functional disorders, and as controls in experiments involving other cell types.

Epithelial-like Cell Strains

Development of new methods for the isolation and propagation of epithelial cell lines is progressing rapidly at the present time. The process is stimulated by the increase in commercial availability of critical reagents such as specific cytokines, adherence factors, and novel dissociation-promoting techniques. Interested readers may consult methods-oriented works for particulars (see, for example, Hay, 1985; McAteer, 1991; Freshney, 1992).

Epithelial strains have been isolated here using normal and tumor tissues from humans and animals (Table 4). For example, normal bronchial tissues from rhesus and african green monkeys were used to establish epithelial strains ATCC CCL 208 (4MBr-5) and ATCC CRL 1576 (12MBr-6), respectively. Glass chips are utilized to select for the epithelial component in cell outgrowths. Chips 2–3 mm in diameter (purchased or prepared by fragmenting coverslips) are washed, sterilized, and placed in 35-mm petri plates. Biopsies of about 3 mm^3 are placed into the petri plates in 0.5 ml of F12K medium supplemented with 15% fetal bovine serum and epidermal growth factor (10–30 ng/ml). Chips are selected that support outgrowths of epithelia free of fibroblast contamination. These are transferred by fine-tipped tweezers to new plates with the same growth medium plus irradiated lung fibroblasts (ATCC CCL 171-MRC5). The resulting epithelial populations are used to establish the cell strains. Cells from these cultures have periodic acid–Schiff (PAS)-positive inclusions and require epidermal growth factor for optimal propagation (Caputo et al., 1979).

Thompson *et al.* (1985) devised and utilized a technique for serum-free cultivation of colonic cells to isolate an epithelial-like strain from female fetal tissue. The cells of ATCC CRL 1790 (CCD841CoN) are propagated in ACL-4 medium in vessels coated with bovine serum albumin, fibronectin, and collagen. The morphology is similar to that of other colonic mucosal lines and, like those, CCD841CoN cells do not contain keratin. Thus definitive evidence for epithelial character is lacking. The cultured cells may have lost ability to synthesize keratin or may represent a primordial stem cell component. The cells are diploid and no consistent marker chromosomes have been observed.

An extremely interesting and useful epithelial-like cell strain, ATCC CRL 1492 (AR42J), was isolated at the ATCC from an exocrine rat pancreatic tumor induced by azaserine (Jessop and Hay, 1980). When grown in the presence of dexamethasone or other steroids, AR42J cells contain pleiomorphic zymogen droplets and significant amounts of amylase and its mRNA, and secrete the enzyme in response to cholecystokinin (Logsdon *et al.*, 1985). Under serum-free culture conditions, AR42J cells exhibit a proliferative response to the peptide hormone gastrin as measured by [^{3}H]thymidine incorporation (Seva *et al.*, 1990).

Hybridomas

The hybridoma bank (HB) of the ATCC began operation in October, 1980. A number of hybridomas producing monoclonal antibodies to cell surface antigens have also been developed and characterized in the ATCC laboratories. Epithelial-specific antigens on human foreskin keratinocytes can be located, for example, using monoclonal Ep-16 from ATCC HB 155. This hybridoma was developed by A. Hamburger *et al.* (1985a) after immunization of BALB/c mice with membrane-enriched preparations from primary cultures of human keratinocytes. NS-1 cells from the tumor immunology bank (TIB) (ATCC TIB 18) were used as fusion partners, and a modified enzyme-linked immunosorbent assay (ELISA) screening method was used to identify positive clones. The antibody reacts very strongly with stratified squamous epithelia. The cell surface antigen detected is distinct from other epithelial-specific antigens as shown by differences in tissue distribution and cellular localization.

Using a similar technique, Hamburger *et al.* (1985b) were also able to develop hybridomas secreting monoclonal antibodies to human umbilical vein endothelia. In this case, intact primary-passage cultured endothelia were used as immunogens and P3X63Ag8.653, a nonsecreting mouse myeloma line (ATCC CRL 1580), was the fusion partner.

Of the various hybridoma lines developed, one designated 14E5 (from ATCC HB 174) yields an antibody that reacts with macrophages, but not fibroblasts or bovine endothelia, whereas another, 12C6 (ATCC HB 173), reacts with human and primate fibroblasts and endothelia from bovine arteries, but not with mature macrophages. A third clone, 10B9 (ATCC HB 172), reacts only with human endothelia and immature macrophages (Table 4). These antibodies should prove useful to detect differentiation markers of endothelia and in studies of endothelial cell function.

Human B Lymphoblastic Lines

Human B lymphocytic cells, once successfully transformed by Epstein–Barr virus (EBV), are continuous and may be used to provide a uniform supply of donor DNA for genetic and other analyses. The method for EBV transformation of fresh or cryopreserved human B lymphocytes as modified and applied here is extremely efficient, giving a success rate greater that 95% (Caputo *et al.*, 1991). Lymphocytes are separated from whole blood by centrifugation through Ficoll. They are transformed by exposure to Epstein–Barr virus obtained as a supernatant from ATCC CRL 1612 (B95-8) cells. Virus production is verified in advance by immunoperoxidase staining after application of monoclonal antibody to EBV capsid antigen (ATCC HB 168) (72A1). The addition of irradiated feeder cells (ATCC CCL 171) (MRC-5) is important to enhance efficiency in lymphoblast culture initiation.

Literally hundreds of human lymphoblastic lines have been developed in our laboratories using this procedure over the past 3 years.

Cryopreservation

Cryopreservation techniques for a large variety of tissue cell types are now well developed, and they have been applied extensively. The utility of such procedures is broad, including, for example, applications, such as transfusion, transplantation, artificial insemination, and conservation of genetic resources, (Ashwood-Smith, 1980). Fairly standard methods have been found satisfactory for most metazoan cell lines, and have been in use since the early 1960s (Freshney, 1987).

In general cells are suspended at $(1–5) \times 10^6$/ml in their usual growth medium containing 5–10% dimethyl sulfoxide (see below). The suspension is dispensed to glass or plastic ampules (1 ml/vial) and frozen by cooling at 1°C/min to about −50°C. This slow freezing step is critical. The ampules are placed immediately and quickly into liquid nitrogen (−196°C) or nitrogen vapor (about −135 to −140°C) for storage. For recovery, ampules are

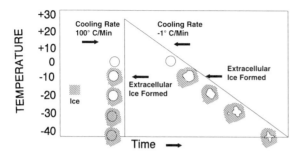

Figure 2. Diagram showing change in cells due to ice crystal formation as a function of freezing rate. Note the high degree of intracellular ice formation in cells cooled rapidly versus those cooled slowly.

removed from the ultracold and are rapidly thawed in a water bath at 37°C. This quick thaw is also very important. Note that it is necessary to wear a face shield and protective clothing whenever handling ampules retrieved from liquid nitrogen (Hay, 1986).

As a physiological suspension or solution is cooled below the freezing point, ice crystals form and the concentration of solutes increases (Figure 2). Freezing slowly (-1°C/min) allows water to escape from the cell by osmosis as the osmotic pressure increases in the fluid in which the cells are suspended. Phase equilibrium studies permit definition of the extent of change in the aqueous phase, and direct observations with cryomicroscopy have documented the effects of such change on viable cells (Leibo et al., 1978; Mazur, 1984). Cryopreservatives such as glycerol (at 10% v/v) and dimethyl sulfoxide (at 5–10%) increase the potential for viable recovery by altering the physical properties of the solutions in which cells are suspended, and by stabilizing cell membranes and proteins (Ashwood-Smith, 1980; Stowell et al., 1965). Factors affecting cell survival during cryopreservation and reconstitution include, for example, water permeability, cell size and cycle stage, cryoprotectant levels and methods of addition, equilibration time, cooling method, type of freezing vessel, thawing technique, and storage temperature (Ashwood-Smith, 1980; Mazur, 1984; Terasima and Yasukawa, 1977). Parameters such as freezing and thawing rates, suspending medium, and protective agents have also been shown to affect preservation of enzyme systems, other proteins, and membrane constituents (Chilson et al., 1965; Pennell, 1965; Tappel, 1966) at least in part through ice crystal formation and osmotic and salt effects. Detailed protocols suitable for the cryopreservation of most mammalian, avian, and insect cell lines have been published elsewhere (Hay, 1978).

Cell Line Banking and Characterization

General descriptions of the seed stock concept and specific, detailed
protocols for cell line characterization have been presented elsewhere (Hay,
1986, 1988; Hay *et al.*, 1989). In brief, key characterizations are performed
on cell line progeny from a seed stock of ampules (Figure 3). These include
tests to assure and quantitate recovery, to assure freedom from microbial
contamination, to confirm specific synthetic function (e.g., immunoglobu-
lins, hormones, and structural proteins), and to verify cell line identity. For
the latter, DNA fingerprinting with probes such as the 33.6, pYNH24, and
others represents a new rapid and efficient technique for identification of
human cell lines (Gilbert *et al.*, 1990; Reid *et al.*, 1990).

The critical need for these tests deserves emphasis because problems
with cell line contamination occur frequently. For example, estimates of
the numbers of lines contaminated with *Mycoplasma* run from about 10%
upward, depending on the laboratories screened and the country of origin.
Results from a testing service in the United States (Table 5) give an incidence of about 10%. *Mycoplasma* contamination can negate interpretation
of virtually any experimental result. Tests for such microorganisms are
essential for any well-controlled facility (Hay *et al.*, 1989).

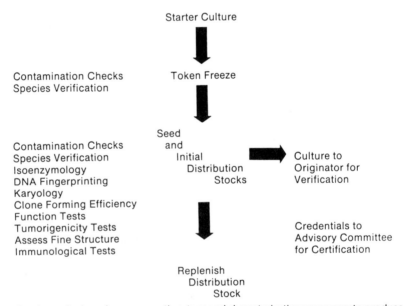

Figure 3. Accessioning scheme presenting steps and characterizations necessary to produce
a cell culture seed stock. The procedure is that used by established cell banks.

TABLE 5
Detection of *Mycoplasma* in Cell Cultures by a Testing
Laboratory 1966–1982[a]

Number of specimens	34,697
Number of positives	3955
Number of strains	4520
Mycoplasmas of human origin (total)	1621
Mycoplasmas of bovine origin (total)	1403
Others	1496

[a] Summarized from Del Giudice and Gardella (1984).

Similarly, the species and identity of cell lines absolutely must be verified because cellular cross contamination is also a major problem. Results from a testing laboratory in the United States, summarized in Table 6, show that overall, 35% of the lines examined were not of the species or precise human identity expected by the inquiring investigators.

By returning to seed stock material to produce each new distribution or replenishment freeze and by repeating key characterizations, the cell bank manager can assure recipients that the cultures supplied are similar both to those provided previously and to those that will be supplied in the future. Research comparability as a function of time and geographical location is thus sustained.

TABLE 6
Summary of Cell Line Cross-Contamination[a]

Reported cell species	Cultures received	Interspecies	Intraspecies	% total
Human	160	40 (25%)	18 (11%)	36
Mouse	27	2	—	8
Cat	24	1	—	4
Others	64	35	—	54
Total	275	78	18	35

[a] Summarized from Hukku *et al.* (1984).

Culture Repositories
National Cell Line Repositories

Argentina

Asociacion Banco Argentino de Celulas (ABAC)
I.N.E.V.H.
C.C. 195
2700 Pergamino
Argentina
Telephone: 54-477-25700
Fax: 54-477-33045
Contact: Dr. Ana Maria Ambrosio
Under development.

China

Kunming Cell Bank of CAS (CBCAS)
Kunming Institute of Zoology
Kunming 650107
China
Telephone: 86-871-82661
Fax: 86-871-82416
Contact: Professor Shi Li-ming
Primarily cell lines from exotic and endangered species.

The Cell Bank of China Center for Type Culture Collection (CBCCTCC)
School of Life Science
Wuhan University
Wuhan, China 430072
Telephone: 86-027-722157
Fax: 86-027-713833
Contact: Associate Prof. Zheng Congyi
Comprehensive.

The Cell Bank of Chinese Academy of Sciences (CBCAS)
Shanghai Institute of Cell Biology
320 Yo-Yang Road, 200031
Shanghai, China
Telephone: 86-4315030, Ext. 41
Fax: 86-4331090
Contact: Professor Ge Xi-rui
Comprehensive.

England

European Collection of Animal Cell Cultures (ECACC)
PHLS Center for Applied Microbiology & Research
Division of Biologics

Porton Down
Salisbury, Wiltshire SP4 OJG
England, United Kingdom
Telephone: 44-980-610391, Ext. 2512
Fax: 44-980-611315
Contact: Dr. Alan Doyle
Comprehensive.

Germany

Deutsche Sammlung von Mikrooganismen und Zellkulturen (DSMZG)
DSM Mascheroder Weg 1 B
D-3300 Braunschweig
Germany
Telephone: 49-531-61870
Fax: 49-531-618750
Contact: Dr. Hans Drexler
Emphasis currently hematopoietic cell lines.

India

National Facility for Animal Tissue and Cell Culture (NFATCC)
Department of Biotechnology
Government of India
Jopasana, Paud Road,
Kothrud, Pune 411 029
India
Telephone: 91-212-335928
Fax: 91-212-369501
Contact: Dr. Ulhas V. Wagh
Comprehensive.

Japan

Institute for Fermentation, Osaka (IFO)
Institute for Fermentation, Osaka
17-85, Juso-honmachi 2-chome
Yodogawa-kn, Osaka 532, Japan
Telephone: 86-300-6555
Fax: 86-300-6814
Contact: Dr. Masao Takeuchi
Comprehensive.

National Institute of Hygienic Sciences (JCRB)
Division of Genetics and Mutagenesis
1-18-1, Kamlyoga, Setagaya-ku,
Tokyo 158, Japan
Telephone: 81-3-3700-1141
Fax: 81-3-3707-6950
Contact: Dr. Hiroshi Mizusawa
Emphasis on lines for cancer research.

Riken Cell Bank (RCB)
3-1 Koyadai, Tsukuba Science City
Ibaraki 305, Japan
Telephone: 81-298-363611
Fax: 81-298-369130
Contact: Dr. Tadao Ohno
Comprehensive.

Korea

Korean Cell Line Research Foundation (KCLRF)
Cancer Research Institute
Seoul Nat'l Univ. College of Medicine
28 Yongun-dong, Chongno-ku
Seoul, 110-744, Korea
Telephone: 82-2-760-3380
Fax: 82-2-742-4727
Contact: Dr. Jae-Gahb Park
Under development. Emphasis on lines for cancer research.

Russia

Russian Cell Culture Collection (RCCC)
Institute of Cytology
Tikhoretsky avenue 4
194064 St. Petersburg
Russia
Telephone: 7-812-247-44-20
Fax: 7-812-247-03-41
Contact: Dr. George Pinaev
Comprehensive.

United States

American Type Culture Collection (ATCC)
Cell Culture Department
12301 Parklawn Drive
Rockville, Maryland 20852
Telephone: 301-231-5529
Fax: 301-231-5551
Contact: Dr. Robert J. Hay
Comprehensive.

Coriell Institute for Medical Research (CIMR)
Coriell Cell Repositories
401 Haddon Avenue
Camden, New Jersey 08103
Telephone: 609-757-4848
Fax: 609-964-0254

Contact: Dr. Richard A. Mullvor
Comprehensive. Emphasis on lines relevant to human genetics and aging.

Acknowledgments

The author is grateful for the skill and assistance of Ms. Shirley Mazur in preparing the figures and the manuscript. We are also pleased to acknowledge support from USPHS Contracts N01-CB-71014 from the NCI and N01-RR-9-2105 from The NIH Center for Research Resources plus Grant 1-R26-CA25635 from The National Large Bowel Cancer Project of The NCI.

References

Adams, R. L. P. (1980). "Cell Culture for Biochemists." Elsevier/North-Holland Biomedical Press, Amsterdam.
Ashwood-Smith, M. J. (1980). *In* "Low Temperature Preservation in Medicine and Biology" (M. J. Ashwood-Smith and J. Farrant, eds.), pp. 19–44. Baltimore Univ. Press, Baltimore.
Butler, M. (1987). "Animal Cell Technology: Principles and Products." Open University Press/ Taylor & Francis, New York/Milton Keynes, UK.
Caputo, J. L., Hay, R. J., and Williams, C. D. (1979). *In Vitro* **15**, 222–223.
Caputo, J. L., Thompson, A., McClintock, P., Reid, Y. A., and Hay, R. J. (1991). *J. Tissue Cult. Methods* **13**, 39–44.
Chilson, O. P., Costello, L. A., and Kaplan, N. O. (1965). *Fed. Proc., Fed. Ann. Soc. Exp. Biol.* **24**, Suppl. 15, S55–S65.
Cunha, A., and Vitkovic, L. (1991). *Tissue Cult. Methods* **13**, 31–38.
Del Giudice, R. A., and Gardella, R. S. (1984). In *Uses and Standardization of Vertebrate Cell Cultures,"* In Vitro Monogr. No. 5, pp. 104–115. Tissue Culture Association, Gaithersburg, MD.
Fraser, M. J. (1989). *In Vitro Cell Dev. Biol.* **25**, 225–235.
Freshney, R. I. (1987). "Culture of Animal Cells: A Manual of Basic Techniques" 2nd ed. Alan R. Liss, New York.
Freshney, R. I. (1992). "The Culture of Epithelial Cells." Wiley, New York.
Gilbert, D. A., Reid, Y. A., Gail, M. H., Pee, D., White, C., Hay, R. J., and O'Brien, S. J. (1990). *Am. J. Hum. Genet.* **47**, 499–514.
Hamburger, A. W., Reid, Y. A., Pelle, B., Milo, G. E., Moyes, I., Krakauer, H., and Fuhrer, J. P. (1985a). *Cancer Res.* **45**, 783–790.
Hamburger, A. W., Reid, Y. V., Pelle, B. A., Breth, L. A., Beg, N., Ryan, U., and Cines, D. B. (1985b). *Tissue Cell* **17**, 451–459.
Hay, R. (1978). *Tissue Cult. Assoc. Manual* **4**, 787–790.
Hay, R. J., ed. (1985). "Methods in Epithelial Cell Culture," J. Tissue Cult. Methods, Vol. 9, No. 2.
Hay, R. J. (1986). *In* "Animal Cell Culture. A Practical Approach" (R. I. Freshney, ed.), pp. 71–112. IRL Press, Washington, DC.
Hay, R. J. (1988). Anal. Biochem. **171**, 225–237.
Hay. R. J. (1991). *In* "Animal Cell Culture and Production of Biologicals" (R. Sasaki and K. Ikura, eds.), pp. 27–39. Kluwer Academic Publishers, Boston.

Hay, R. J., Macy, M. L., and Chen, T. R. (1989). *Nature (London)* **339,** 487–488.

Hay, R. J., Caputo, J., Chen, T. R., Macy, M. L., and Reid, Y. A. (1992). "Catalogue of Cell Lines and Hybridomas," 7th ed. Am. Type Cult. Collect., Rockville, MD.

Hay, R. J., Williams, C. D., Macy, M. L., and Lavappa, K. S. (1982). *Am. Rev. Respir. Dis.* **125,** 222–232.

Hink, W. F. (1979). *In* "Methods in Enzymology" (W. Jakoby and I. Pastan, eds.), Vol. 58, pp. 450–466. Academic Press, New York.

Hink, W. F., ed. (1989). "Invertebrate Cell Cultures," J. Tissue Cult. Methods, Vol. 12, No. 1.

Ho, C. S., and Wang, D. I. C., eds. (1991). "Animal Cell Bioreactors." Butterworth-Heinemann, Boston.

Hukku, B., Halton, D. M., Mally, M., and Peterson, W. D., Jr. (1984). *In* "Eukaryotic Cell Cultures" (R. T. Acton and J. D. Lynn, eds.), pp. 13–31. Plenum, New York.

Jessop, N. W., and Hay, R. J. (1980). *In Vitro* **16,** 212.

Leibo, S. P., McGrath, J. J., and Cravalho, E. G. (1978). *Cryobiology* **15,** 257–271.

Logsdon, C. D., Moessner, J., Williams, J. A., and Goldfine, I. D. (1985). *J. Cell. Biol.* **100,** 1200–1208.

Loo, D. T., Fuquay, J. I., Rawson, C. L., and Barnes, D. W. (1987). *Science* **236,** 200–202.

Lydersen, B. K., ed. (1987). "Large Scale Cell Culture Technology." Hanser Publishers, New York.

Mazur, P. (1984). *Am. J. Physiol.* **247,** C125–C142.

McAteer, J. A., ed. (1991). "Methods in Kidney Epithelial Cell Culture," J. Tissue Cult. Methods, Vol. 13, No. 3.

Pennell, R. B. (1965). *Fed. Proc., Fed. Am. Soc. Exp. Biol.* **24,** Suppl. **15,** S269–S274.

Reid, Y. A., Gilbert, D. A., and O'Brien, S. J. (1990). *ATCC Q. Newsl.* **10**(4), 1–3.

Sasaki, R., and Ikura, K., eds. (1991). "Animal Cell Culture and Production of Biologicals." Kluwer Academic Publishers, Boston.

Schaeffer, W. I. (1984). *In* "Uses and Standardization of Vertebrate Cell Cultures" In Vitro Monogr. No. 5, pp. 19–24. Tissue Culture Association, Gaithersburg, MD.

Seva, C., Scemama, J. L., Bastie, M. J., Pradayrol, L., and Vayasse, N. (1990). *Cancer Res.* **50,** 5829–5833.

Spier, R. E. (1991). *In* "Animal Cell Culture and Production of Biologicals" (R. Sasaki and K. Ikura, eds.), pp. 41–46. Kluwer Academic Publishers, Boston.

Steinsvag, S. K. (1991). *J. Tissue Cult. Methods* **13,** 1–4.

Stowell, R. E., Young, D. E., Arnold, E. A., and Trump, B. F. (1965). *Fed. Proc., Fed. Am. Soc. Exp. Biol.* **24,** Suppl. **15,** S115–S143.

Tappel, A. L. (1966). *In* "Cryobiology" (H. T. Meryman, ed.). pp. 163–177. Academic Press, New York.

Terasima, T., and Yasukawa, M. (1977). *Cryobiology* **14,** 379–381.

Thompson, A. A., Dilworth, S., and Hay, R. J. (1985). *J. Tissue Cult. Methods* **9,** 117–122.

CHAPTER 9

Human and Animal Viruses

Judy A. Beeler

Background

Historical Perspective

Shackell's successful treatment of rabies virus in 1909 represents the first published account of preservation of an animal virus using a process that was an early version of freeze-drying (Shackell, 1909). Prior to this, viruses were maintained by passage in animals. However, freeze-drying provided a way to maintain virally infected material over long periods of time with relative ease when compared to serial passage in a susceptible animal host. The first experiment to demonstrate that this method could provide long-term stability was reported for a bovine virus that had been freeze-dried in 1916 and was shown to be viable after being maintained at room temperature for 30 years (Fasquelle and Barbier, 1950). Later, as mechanical freezers became more available, freezing viruses at −25°C or less was adopted as a practical and reliable method of virus preservation. Influenza virus-infected mouse lung tissue and yellow fever virus were successfully frozen at −76°C by Turner (1938), and Horsfall (1940), using Turner's methods, demonstrated successful storage of influenza, mouse pneumonitis, and canine distemper viruses. Following these early successes, viral repositories, diagnostic and research laboratories, and vaccine manufacturing facilities have relied on freezing and freeze-drying for preservation of viruses and viral vaccines.

The reasons for preserving viruses are similar to those for the preservation of other microorganisms and cells: successful preservation allows for long-term maintenance of consistent stocks and for reproducibility in the testing of viral samples and manufacture of vaccines. For example, field and clinical isolates need to be preserved for transport to testing laboratories; harvested virus from animals or cell cultures may not be processed immediately and require preservation until used. Viruses held to a low number of passages in animals or cell cultures represent a viral population that is similar to that found in nature, and freezing these pools guards against genetic mutations that may occur during subsequent passage. Aliquots of viral stocks frozen at a designated passage level can then be used for multiple and repeatable experiments with the same viral population. Furthermore, it is extremely important that consistency be maintained during the production of viral vaccines; new lots of final product can be prepared with frozen viral seed stocks that consistently reproduce the desired immunogenic and attenuation characteristics. For example, the World Health Organization (WHO) maintains aliquots of the original attenuated poliovirus strains developed by Dr. Albert Sabin in the 1950s. These aliquots are frozen and, when amplified properly and kept to the prescribed low passage level, should produce a vaccine today with the same properties that characterized the original Sabin vaccine. Similarly, viruses that are currently under development for use as vectors for gene therapy must be highly characterized and homogeneous for safe and effective use and need to be maintained as frozen stocks.

In order to better appreciate the requirements for freezing and freeze-drying of human and animal viruses, some consideration must first be given to understanding the structural and functional organization of this diverse group of microorganisms. The classification of viruses is based on morphological and physiochemical properties. Thus viruses are divided into those with DNA or RNA genomes and further subdivided into families based on size and structural properties. There are 6 families of DNA viruses and 13 families of RNA viruses (see Table 1). The genomes of viruses within a given family are similar in terms of their polarity, genome organization, and method of replication and have morphologic features in common, including size, shape, nucleocapsid symmetry, and presence or absence of an envelope. Other characteristics that help to classify viruses include sensitivity to low pH, heat, and lipid solvents. Sensitivity to pH is determined by subjecting a virus to a pH of 3.0. Loss of titer greater than one $\log_{10}$ indicates that the virus is pH sensitive and acid labile. Heat treatment of a virus for 30 min at 50°C is a test of thermal stability. If the virus survives this treatment, with no more than a one $\log_{10}$ loss of infectivity, it is considered to be thermostable. Treatment with lipid solvents, commonly ether and

TABLE 1
Properties of the Families of Viruses

Family	Example	Genome[a]	Virion Shape	Virion Size (nm)
DNA viruses				
Hepadnaviridae	Hepatitis B virus	ds, circular	Spherical	42
Papoviridae	Papillomavirus	ds, circular	Icosohedral (72)[b]	45–55
Adenoviridae	Adenovirus	ds, linear	Icosohedral (252)	70–90
Herpesviridae		ds, linear	Icosohedral (162), enveloped	120–200 (envelopes), 100–110 (capsid)
	Herpes simplex, EBV			
Poxviridae	Vaccinia	ds, linear	Brick-shaped, enveloped	300 × 240 × 100
Parvoviridae	Parvovirus B19	ss, linear	Icosohedral (32)	18–26
RNA viruses				
Picornaviridae	Poliovirus	ss, 1, +	Icosohedral	25–30
Caliciviridae	Norwalk virus?	ss, 1, +	Icosohedral	35–40
Togaviridae	Rubella	ss, 1, +	Spherical, envelope	60–70
Flaviviridae	Dengue virus	ss, 1, +	Spherical, envelope	40–50
Orthomyxoviridae	Influenza virus	ss, 8, −	Spherical envelope	80–120
Paramyxoviridae	Measles virus	ss, 1,−	Spherical, envelope	150–300
Coronaviridae	Coronavirus	ss, 1, +	Spherical, envelope	60–220
Arenaviridae	Lassa fever virus	ss, 2, −	Spherical, envelope	50–300
Bunyaviridae	Hantaan virus	ss, 3, −	Spherical, envelope	90–100
Retroviridae	HIV-1	ss, 1, +	Spherical, envelope	80–110
Rhabdoviridae	Rabies virus	ss, 1, −	Bullet-shaped, envelope	180 × 75
Filoviridae	Ebola virus	ss, 1, −	Filamentous, envelope	800× 80
Reoviridae	Rotavirus	ds, 10–12	Icosohedral	60–80

[a] ds, Double stranded; ss, single stranded; +, positive stranded RNA genome; −, negative stranded RNA genome.
[b] Number of capsomeres per virion.

chloroform, determines whether a virus has a lipid-containing envelope. Despite the similarities among the physical properties of these families of viruses, relatively few general rules for freezing or freeze-drying viruses can be made. For example, it is common to use membrane-stabilizing agents to preserve viruses with a lipid envelope. Picornaviruses, which lack a lipid membrane, maintain infectivity in the presence of $MgCl_2$. However, the optimal conditions for freezing specific viruses need to be determined empirically.

Industrial Importance

Significant reduction of morbidity and mortality for many viral infections has been accomplished by the successful application of viral vaccines. In the case of smallpox, eradication of this important human pathogen has been realized. Polio has now been targeted as the next viral disease that can be eradicated by immunization. Vaccine production requires the use of the "seed lot system" for consistent reproduction and quality control of vaccine lots. Vaccine virus seeds are usually frozen at two passage levels prior to vaccine lot production and these stocks of virus are designated the master seed and the production or working seed, respectively. Multiple aliquots of these seeds are prepared for reference as well as for characterization and standardized testing for possible contaminants. Finally, it is also important to distribute the storage of the master virus seed and working seed among multiple freezers in order to be assured of preserving part of the stock in the event of a mechanical failure. The seed lot system establishes a viral stock that can be standardized and provides a source of viruses so that vaccine lots manufactured over decades will be closely related to the original seed lot. This is especially important for live-attenuated vaccines that in most cases have been derived by multiple serial passage of a clinical isolate of a virus. Deviation from the attenuated passage number may result in undesirable phenotypic changes of the attenuated vaccine virus. This was the case with yellow fever vaccine when uncontrolled passage of this virus resulted in a vaccine with a higher than normal incidence of side reactions (Fox *et al.*, 1942).

Presently, researchers are using the tools of molecular biology to produce a new class of recombinant viral agents that will be used for immunization or genetic therapy. For example, mammalian retroviruses that integrate into the host cell genome are being developed as vectors for delivery of human DNA sequences. The newly acquired DNA is expressed in the form of protein(s) that may be deficient in the host cell or may require *de novo* synthesis. This form of genetic therapy has been demonstrated routinely *in vitro* and *in vivo* in animal experiments and is being proposed as therapy for some human genetic diseases (for review, see Roemer and Friedmann, 1992). In addition, recombinant viral vectors are also paving the way for a novel type of therapy in which cancer cells can be specifically targeted for infection by a recombinant virus; following infection the inserted gene is expressed and the gene product selectively makes the infected cancer cells more susceptible to subsequent chemotherapy or to natural immune surveillance in the host (Collins et al., 1992; Culver et al., 1992). Alternatively, recombinant vectors may express antisense RNA that can bind to and block the activity of the host gene sequences. Successful experiments of this kind require viral vectors that are thoroughly characterized and safe. Seed stocks of the recombinant viral vectors constructed to express the desired nonviral gene sequences must be frozen or freeze-dried in order to be preserved for future experiments and clinical trials.

A somewhat similar approach is being developed for vaccination using viral vectors as delivery vehicles for genes important for immunization. For example, one approach uses a highly attenuated vaccinia virus that carries, as part of its genome, various immunogenic proteins that would evoke protection against infection (Moss and Flexner, 1987). Other viral vectors used similarly include adenovirus and canarypox.

Antigens used for viral diagnostic purposes must also be quality controlled and, in general, would also be produced from a virus seed lot system. Variation of quality or quantity of viral antigens in diagnostic kits and reagents may directly influence results in the diagnostic laboratory. Regulation of the manufacture of diagnostics is similar to those for vaccine manufacture. Regulations concerning the manufacture of biological products are published by the United States Public Health Service (21 CFR 600) and outline the requirements for control of reagents and raw materials in the manufacture of diagnostics to screen blood and blood products. For these purposes, it is essential to use viral seed stocks that are appropriately characterized and preserved by freezing.

Characterization

Viability of viruses is usually determined by the use of infectivity assays. These include assays whereby viral cytopathology is measured in cell cultures or, when this is not possible, potency may be determined by inoculation of animals or eggs. By far the most accepted and accurate quantitative measure of viral infectivity is the plaque assay. Most viruses form plaques in susceptible host cells. Generally, after virus inoculation on suitably sensitive cell monolayers, an agar overlay is applied so that viral cytopathology is localized. After an incubation period, viable cells are stained with a vital dye, revealing areas of cell death, the plaque-forming unit. Most viral freezing and recovery experiments use the plaque assay to titer virus before and after treatment as a measure of viral activity. In this way, one can determine the efficiency of the freezing or freeze-drying process. Electron microscopy (EM) has also been used on occasion to visualize the effects of freezing and freeze-drying. These EM experiments are limited by the purity and titer of the viral preparation as well as the subjective interpretation of observations.

Methods for Preservation

Harvesting and Preparing Viruses

The success of freezing or freeze-drying of viruses is dependent on the proper preparation of the harvested virus. The source of virus may be animal tissues or organs, animal sera, virus-infected cell cultures, and, on occasion, egg allantoic or amniotic fluids.

The harvested tissue will depend on the particular host-range of the virus being grown. Harvesting methods are also dependent on the properties of viral replication. Some viruses, e.g., herpes viruses, stay cell associated whereas other are readily released from the cell into the culture fluids. In general, viruses grown after passage *in vivo* require disruption of the infected animal tissue or organ by homogenizers or aerosol-controlled blenders. Diluent consisting of buffer or cell culture media is added for release of the virus into the aqueous phase during cell disruption. Depending on the virus, serum may be required as an additive to the diluent for optimal freezing (Table 2). Viruses that are found in blood may be harvested and frozen directly in the serum after separation from blood cells or by cocultivating infected peripheral blood lymphocytes (PBLs) with susceptible cell lines. These preparations may be frozen without further treatment, although it is generally recommended that serum or PBLs be frozen and thawed only once to achieve optimal virus recovery. Similarly, cell culture harvests may require disruption of the cell monolayer as a first step. Cells may be scraped from the culture vessel surface into the culture fluids or these fluids may be removed and the cells treated separately. In either case, cells are disrupted and virus is released. In certain cases, virus-infected cells are harvested for a seed preparation. Gentle scraping and harvesting of these cells into the culture fluids or into fresh buffer or cell culture medium is required. The specific details for optimal harvesting procedures for individual viruses may be found in the literature.

Usually animal serum is used as a component of cell culture medium. Fetal bovine serum (FBS) is the most frequently used additive by cell

TABLE 2
Optimal Conditions for Freezing and/or Freeze-Drying Some Mammalian Viruses

Virus	Stabilizers	Storage temperature	References
Herpes	Serum, 7% DMSO in culture medium	−70° frozen	Wallis and Melnick (1968)
Herpes	Sodium glutamate	−20°C freeze-dried	Scott and Woodside (1976)
Varicella-zoster	Sucrose, sodium glutamate, albumin, KPO_4	−70°C frozen	Grose (1981)
Dengue	20–50% fetal bovine serum, 0.75–2.0% bovine albumin	−70° frozen	Shope and Sather (1979)
Sindbis	20–50% fetal bovine serum, 0.75–2.0% bovine albumin	−70°C frozen	Shope and Sather (1979)
Measles	Serum, DMSO	−70°C frozen	Wallis and Melnick (1968)
Vesicular Stomatitis	Serum, DMSO	−70°C frozen	Wallis and Melnick (1968)
Respiratory syncytial	$MgSo_4$ and HEPES buffer	−70° C frozen	Fernie and Gerin (1980)
Adenovirus	Culture medium	−70°C frozen	Wallis and Melnick (1968)
Poliovirus	Culture medium	−70°C frozen	Grieff and Rightsel (1967)
Poliovirus	Dialyze out salts	−20°C freeze-dried	Berge et al. (1971)
Vaccinia	Culture medium	−70°C frozen	Wallis and Melnick (1968)
Influenza	Serum	−70°C frozen	Dowdle et al. (1979)
Influenza	0.5% gelatin	−20–4°C freeze-dried	Dowdle et al. (1979)
Parainfluenza	0.5% bovine albumin, serum	−60°C or greater	Chanock (1979)
Mumps	Chicken amniotic fluid	−70°C	Hopps and Parkman (1979)
Rabies	0.75% bovine albumin, serum	−70°C	Johnson (1979)
Hepatitis A	2–40% stool suspension	−70°C	Feinstone et al. (1979)
Hepatitis A	Culture medium	−70°C	Feinstone et al. (1979)

culturists and virologists because it is nontoxic to most cells and the protective effect during freezing is consistent. Therefore, FBS doubles for cell nutrition as well as a cryoprotectant for freezing and freeze-drying. Concentrations used for cell culturing and growing viruses range from 0.5 to 5% in cell culture medium. In many cases the cell culture medium containing serum constitutes the cyroprotective diluent that will be used for freezing. However, further addition of serum is required for many viruses (see Table 2). Infectious materials should be handled according to the biosafety guidelines that apply for each virus.

Freezing

TEMPERATURE

The rate of freezing and the final temperature must be considered for virus preparations being frozen for long-term storage. To maintain viral infectivity throughout this process, the integrity of the viral capsid, the viral envelope (if there is one), and the viral nucleic acid must be preserved. Dimmock (1967) demonstrated that viral infectivity may be reduced by heat damage to both viral proteins and nucleic acid. He showed that the stability of these components varies with elevated temperature so that inactivation at a particular temperature takes place through whichever component is least stable at that temperature. In general, most viruses may be "snap frozen" in a dry ice/ethanol bath prior to storage at temperatures $\leq -70°C$. It may be preferable to freeze some viruses, such as cytomegalovirus or varicella-zoster virus, as viable infected cells. In this case, a slow controlled rate of freezing is recommended: cooling should be done at 1°C/min to$-40°C$ and then at 10°C/min to $-90°C$ (Simione and Brown, 1991). It has also been reported that influenza and measles virus may benefit from this two-step freezing process (Rowe and Snowman, 1976).

Storage temperature is also critical for some viruses with envelopes, such as influenza virus, respiratory syncytial virus, and measles virus. These all require storage at $-65°C$ or lower for optimal recovery after long-term storage (Rightsel and Greiff, 1967; Greiff et al., 1964; Law and Hull, 1968).

It is common practice for today's virologist to store frozen viruses in ultralow-temperature freezers at temperatures of -65 to $-85°C$. Freezing in liquid nitrogen or its vapor phase, a common practice for preservation of mammalian cells, is not necessary for most viruses. Liquid nitrogen storage may be used as a "backup" system for viral seed repositories. No known conventional virus requires $-196°C$ (liquid) or $-150°C$ (vapor) nitrogen for short- or long-term storage.

CRYOPROTECTANTS

In one of the first comprehensive studies of freezing and freeze-drying with representative members of most of the RNA and DNA virus families,

Rightsel and Greiff (1967) demonstrated the importance of the suspending medium used for freezing and freeze-drying. Medium containing skim milk, calcium lactobionate, normal serum albumin, dimethyl sulfoxide (DMSO), glycerol, or magnesium chloride promoted better freezing and freeze-drying recovery with a range of viruses, especially enveloped RNA viruses. Wallis and Melnick (1968) further demonstrated that DMSO or serum acted as a cryoprotectant for four different enveloped viruses, herpes virus, measles virus, Sindbis virus, and vesicular stomatitis virus. Under the conditions described DMSO performed somewhat better than serum as a cryoprotectant. The nonenveloped viruses, vaccinia, adenovirus, and poliovirus, did not require DMSO or serum for retention of viability during freezing. The authors concluded that enveloped viruses were similar to mammalian cells and that DMSO stabilizes the viral envelope much as it stabilizes the mammalian plasma membrane. However, another member of the herpes virus family, varicella-zoster (VZ), is especially sensitive to freezing and freeze-drying. Grose et al., (1981) used sucrose, potassium phosphate, sodium glutamate, and albumin in a cryoprotective "cocktail" for VZ virus. For both cell-associated and cell-free virus, the cocktail was fully protective, with 100% recovery of infectivity after freezing to $-70°C$ and freeze-drying. Additionally, these authors found that glutamate and albumin could be removed from the cocktail without loss of protection. In contrast, Scott and Woodside (1976) showed that sodium glutamate was essential for the stability of freeze-dried herpes virus.

Certain viruses are stabilized by the addition of divalent cations. For example, respiratory syncytial virus was found to be stable for up to 1 month at 4°C following the addition of magnesium sulfate, and the titer of virus was maintained through several cycles of freezing and thawing when $MgSO_4$ was added (Fernie and Gerin, 1980). Likewise, poliovirus was stabilized in the presence of $MgCl_2$ (Melnick et al., 1963). Alternatively, 300 mM sodium phosphate at pH 6.7 stabilized poliovirus at 0–6°C for up to 12 months (Mauler and Gruschkau, 1978).

Cryoprotectants and the rate of freezing minimize the formation of ice crystals that can damage the viruses; similarly, it is common practice to rapidly thaw frozen viral stocks at a previously determined optimal temperature to minimize ice crystal formation.

Freeze-Drying

FREEZE-DRYING CYCLES

Freeze-drying, or lyophilization, is a method for preserving viruses and other biological materials by removing water from frozen samples via sublimation. The result is a dried preparation that is usually more stable than wet-frozen preparations at various temperatures and for longer periods. The process is usually divided into three stages: prefreezing, sublimation drying under vacuum (primary drying), and desorption or secondary drying. Prefreezing can be accomplished in special freezers or in the freeze-dryer itself. A freeze-dryer has a vaccuum pump that removes air from a chamber that holds vials of the material to be freeze-dried. Water vapor generated through this process is condensed into a refrigerated trap and is not allowed to enter the vacuum pump. The rate of sublimation can be controlled by regulating the heat transfer to the frozen material. After final drying, inert gas is used to fill the vials and protect the freeze-dried material from the deleterious effects of oxygen. The freeze-drying cycle requires optimization for each virus.

A typical cycle that has been used successfully for dengue live-attenuated virus vaccines (all four serotypes) consisted of freezing to $-40°C$ for 2 hours and allowing the condenser to reach a temperature of $-60°C$ (K. Eckels, unpublished data). Vacuum was drawn to 100 μm of Hg, at which time heat was applied at a controlled rate overnight (approximately 18 hours). Final drying occurred at $20–23°C$ for 2 hours followed by backfilling the vials with dry, sterile nitrogen and capping. Vaccine vials were finally stored at $-30°C$.

Another cycle that preserves viral infectivity has been used for dengue candidate vaccines; this consisted of freezing to $-60°C$ overnight followed by raising the temperature to $-20, 0$, and finally to $10°C$ over a 2-day period while vaccuum was applied (K. Eckels, unpublished data). In both cases, care was taken to avoid subjecting this thermolabile virus to prolonged periods of high temperatures that could result in inactivation of viral infectivity. Greiff and Rightsel (1967) have successfully freeze-dried influenza and measles virus with cycles extending over a 48-hour period with shelf and product temperatures not exceeding $0°C$. Greiff (1991) has outlined an approach for the development of a successful freeze-drying cycle. Similarly, by varying drying time, shelf temperature, and vacuum pressure an optimal lyophilization cycle for varicella-zoster vaccine was determined (Bennett *et al.*, 1991). Freeze-drying methods and cycles used by the American Type Culture Collection (ATCC) for their virus stocks can be found in Simione and Brown (1991).

STABILIZERS AND ADDITIVES

Stabilizers added to the viral suspension medium for freeze-drying are many of the same used for freezing. These additives promote preservation of viral infectivity by acting as antioxidants and by providing "bulk" to the virus preparation to allow easier reconstitution. Also, they should be nonhygroscopic to avoid moisture contamination of the freeze-dried material. For vaccines, the stabilizers must also be nonimmunogenic and nonreactive in the vaccine recipient. One of the first stabilizers used for viral freeze-drying was 2.5% gum acacia to preserve the infectivity of vaccinia virus (Rivers and Ward, 1935). Peptone and normal horse serum were also successfully used for freeze-drying of vaccinia virus (Collier, 1951; Sparkes and Fenje, 1972). Calcium lactobionate and human serum albumin were cryoprotective for measles virus during freeze-drying (Greiff and Rightsel, 1967), whereas cell culture medium alone was less protective. A combination of sucrose, phosphate, sodium glutamate, and albumin was used for freeze-drying pseudorabies and herpes viruses (Scott and Woodside, 1976; Calnek *et al.*, 1970). Varicella-zoster virus was also freeze-dried successfully in this combination of stabilizers, but Grose (1981) found that he could eliminate glutamate and albumin from the mixture and still retain 100% recovery of this very labile virus. DMSO cannot be used for stabilization in freeze-dried preparations because it becomes concentrated to toxic levels (Greiff and Rightsel, 1967).

A group of viruses that requires special attention to stabilizing media is the enteroviruses, for example, poliovirus and hepatitis A virus. Early work with poliovirus demonstrated that it had a high degree of lability when freeze-dried, even with stabilizers. The lability is due to the presence of inorganic salts that may be present in the culture medium. When these are removed by dialysis prior to freeze-drying, losses do not occur (Berge *et al.*, 1971). The ATCC uses ultrafiltration, a more rapid type of dialysis to remove salts prior to freeze-drying of enteroviruses (Simione and Brown, 1991).

Vaccines that require freeze-drying have special stabilizer requirements that allow them to be used safely in humans. Animal serum cannot be used as a stabilizer and must be completely removed from the vaccine prior to bottling. When a protein substitute is required, human serum albumin has been used successfully for many vaccines. More recently, hydrolyzed gelatin has been used as a replacement for albumin (Table 3). Both of these stabilizers are acceptable for human vaccines because they usually do not stimulate deleterious immune responses in the vaccinee. Sugars such as sucrose, lactose, and sorbitol are also used either alone or in combination with a protein stabilizer. Cell culture medium is often used as a buffer because many live vaccines are harvested in their culture medium without further processing.

MOISTURE CONTENT

Optimal moisture content of the freeze-dried product needs to be empirically determined for each viral preparation. A good freeze-drying cycle can attain a moisture content of 1% or less in the final product.

TABLE 3
Optimal Conditions for Stabilization and Storage of Licensed and Experimental Viral Vaccines

Vaccine	Stabilization	Storage
Smallpox (Dryvax; Wyeth Laboratories)	No stabilizers added to the freeze-dried calf lymph	2–8°C
Measles (Attenuvax; Merck, Sharp, and Dohme)	Sorbitol, hydrolyzed gelatin	2–8°C
Mumps (Mumpsvax; Merck, Sharp, and Dohme)	Sorbitol, hydrolyzed gelatin	2–8°C
Rubella (Meruvax II; Merck, Sharp, and Dohme)	Sorbitol, hydrolyzed galatin	2–8°C
Polio, live oral (Orimmune; Lederle Laboratories)	Sorbitol stabilizer used in this non-freeze-dried vaccine	<0°C (frozen)
Yellow fever (YF-Vax; Squibb/Connaught)	Sorbitol, gelatin	0°C or less (frozen)
Dengue (experimental)	Lactose, human serum albumin, or hydrolyzed gelatin	−20°C or less (frozen)
Rabies (Imovax; Institut Merieux)	No stabilizer used	2–8°C
Influenza(Fluzone; Squibb/Connaught)	No stabilizer used in this formalin-inactivated, non-freeze-dried vaccine	2–8°C

However, depending on the virus, a higher moisture content may be desirable. A varicella virus vaccine with 6–8% moisture following freeze-drying was found to be more stable than vaccines containing less moisture (Bennett *et al.*, 1991). Various viruses have been freeze-dried to approximately 1% moisture and shown to be stable over periods of time at 4°C, but better long-term stability can be achieved by storage at −20°C.

STABILITY

Assays for retained infectivity following freeze-drying are done on rehydrated vials of freeze-dried virus. A volume of sterile, distilled water is used to reconstitute the vial, with titration of the virus done immediately in the appropriate plaque or infectivity assay. The best control for baseline infectivity would be the virus harvest that was assayed prior to freeze-drying. Often, frozen and thawed specimens are used for baseline titers. Freeze-dried viruses are normally stored at 4°C or −20°C. Accelerated stability studies can be done to test thermostability for shorter periods at elevated temperatures. Often an ambient temperature of 25°C or temperatures of 35–37°C are used for accelerated studies. These temperatures are chosen to study thermostability for viruses, mainly vaccines, that may not have continuous refrigeration available up to the time of use. This is a problem for many live viral vaccines that are being used in developing countries (Widdus *et al.*, 1989).

The largest application of freeze-drying of viruses occur for live-attenuated viral vaccines that are thermolabile. The first viral vaccine to be freeze-dried was yellow fever vaccine (Penna, 1956). Recent advances have resulted in more stable, freeze-dried yellow fever vaccines (Robin *et al.*, 1971; Burfoot *et al.*, 1977). The WHO standard for stability of this vaccine is no more than one $\log_{10}$ loss of potency held at 37°C for 2 weeks. A recent collaborative study of yellow fever vaccines from 12 manufacturers demonstrated that 7 of 12 vaccines met this requirement. Similar WHO stability standards are in place for live, freeze-dried measles vaccine. A method to freeze-dry and stabilize trivalent, live, attenuated, oral polio vaccine is being sought so that this vaccine can be delivered and used in developing countries without significant loss in potency. It has been proposed that thermoresistant mutants of polioviruses might serve as the prototype for the development of heat-stable vaccine strains (Kew, 1989). Alternatively, organic compounds such as WIN 51711 and R 78206, which bind to polivirus VP1, have been found to increase the half-life of poliovirus antigen 50- to 250-fold at temperatures up to 48°C (Rombaut *et al.*, 1991). These drugs stabilize the conformation of the viral capsid and exhibit a potent antiviral effect that would not make them suitable as stabilizers for a live virus vaccine. In the future, similar compounds could be designed to stabilize but not inhibit the replication of live poliovirus strains. Other licensed, freeze-dried live-viral vaccines are those for mumps and rubella, whereas vaccines for varicella (chicken pox) and cytomegaloviruses are being developed that will require stabilization by freeze-drying. Table 3 lists licensed vaccines as well as those still being tested and the available published data on freeze-drying of these vaccines.

Virus Repositories

American Type Culture Collection
12301 Parklawn Drive
Rockville, Maryland 20852
Telephone: 1-800-638-6597 (United States and Canada, only) or 1-301-881-2600
FAX: 1-301-231-5826
Request for international reference materials should be made with a statement of intent from the investigator.

International Laboratories for Biological Standards

National Institute for Biological Standards and Control
Blanch Lane, South Mimms
Potters Bar, Hertsforshire EN6 3QG
England
Telephone: 707-54753/54763
FAX: 707-46730
Contact: Dr. Timothy Forsey

Staten Seruminstitut
80 Amager Boulevard
2300 Copenhagen Street
Copenhagen
Denmark
Telephone: 45-32-95-28-17 or 45-32-68-34-66
FAX: 45-32-68-38-68 or 45-32-68-31-50
Contact: Dr. Jorn Lyng or Gert Albert Hansen

Other Laboratories

Anti-Viral Research Branch
National Institutes of Allergy and Infectious Disease
National Institute of Health
Bethesda, Maryland 20892
Telephone: 301-496-8285
Contact: Thelma Gaither for catalogue of available reagents

Centers for Disease Control

Atlanta, Georgia 30333
Telephone: 301-404-639-3311/3355
FAX: 301-404-639-3037/3296

Rijksinstituut voor Volksgezondheid en Milieuhygeine Postbus 1

3720 BA
Bilthoven, The Netherlands
Telephone: 30-749111
FAX: 30-742971

Influenza Viruses and Reagents

Commonwealth Serum Laboratories
45 Poplar Road
Parkville, Victoria
Australia 3052
Telephone: 61 3 389 1340
FAX: 61 3 388 2063
Contact: Alan W. Hampson

National Institute for Biological Standards and Control
Blanche Lane, South Mims
Potters Bar, Herts. EN6 3QG
England
Contact: Dr. John Woods

WHO Collaborating Center for Influenza
Centers for Disease Control
Influenza Branch G-16
1600 Clifton Road
Atlanta, Georgia 30333
Telephone: 404-639-3591
FAX: 404-639-2334
Contact: Dr. Nancy Cox

Division of Virology
HFM 463
Center for Biologics Evaluation and Research
Food and Drug Administration
29A/1D10
8800 Rockville Pike
Bethesda, Maryland 20892
Telephone: 301-496-6828
FAX: 301-496-1810
Contact: Michael Williams or Roland Levandowski, MD

HIV

AIDS Research and Reference Reagent Program
National Institute of Allergy and Infectious Disease
National Institutes of Health
Bethesda, Maryland 20892
Telephone: 301-340-0245

FAX: 310-340-9245
Contact: Ogden Bioservices Corp.

Arbovirus Collections

Centers for Disease Control
Division of Vector Borne Infectious Diseases
P.O. Box 287
Fort Collins, Colorado 80525
Telephone: 970-221-6425
Contact: Dr. Nick Karabatos

Department of Pathology
University of Texas Medical Branch
Galveston, Texas 77555
Telephone: 409-772-6662
Contact: Dr. Robert Shope

Epidemiology and Public Health
Yale University School of Medicine
New Haven Connecticut 06520
Telephone: 203-785-6976
Contact: Dr. Rebecca Rico-Hesse

Polioviruses/WHO Repository

ATCC for Division of Viral Products
Center for Biologics Evaluation and Research
Food and Drug Administration
8800 Rockville Pike
Bethesda, Maryland 20892
Telephone: 301-496-5041/301-443-8411
FAX: 301-496-1810
Contact: Dr. Ronald Lundquist or Dr. Jacqueline Muller

Respiratory Syncytial Virus

WHO Reagent Bank for RSV and PIV3
Division of Viral Products
Center for Biologics Evaluation and Research
Food and Drug Administration
8800 Rockville Pike
Bethesda, Maryland 20892
Telephone: 301-402-0414/0415
FAX: 301-496-1810
Contact: Judy Beeler, MD

Measles Virus

WHO Reagent Bank for Measles Virus
Insitut Pasteur de Lyon

Unite d'Immunologie et Stategie Vaccinale
Avenue Tony-Garnier
69365
Lyon, Cedex 97
France
Contact: Dr. T. Fabian Wilde

Acknowledgments

The author would like to thank Dr. Kenneth Eckels for guidance and sharing data on freeze-drying dengue viruses and Drs. Karen Goldenthal, Ron Lundquist, and Hana Golding for reviewing the manuscript.

References

American Type Culture Collection (ATCC) (1990). "The Catalogue of Animal Viruses and Antisera, Chlamydiae, and Rickittsiae," 6th ed. ATCC, Rockville, MD.
American Type Culture Collection (ATTC) (1991). "ATCC Preservation Methods: Freezing and Freeze-drying," 2nd ed. ATCC, Rockville, MD.
Bennett, P. S., Maigetter, R. Z., Olson, M. G., Provost, P. J., Scattergood, E. M., and Schofield, T. L. (1991). *Dev. Bio. Stand.* **74,** 215–221.
Berge, T. O., Jewett, R. L., and Blair, W. O. (1971). *Appl. Microbiol.* **16,** 362–365.
Burfoot, C., Young, P. A., and Finter, N. B. (1977). *J. Biol. Stand.* **5,** 173–179.
Calnek, B. W., Hitchner, S. B., and Adldinger, H. K. (1970). *Appl. Microbiol.* **20,** 723–726.
Chanock, R. M. (1979). *In* "Diagnostic Procedures for Viral, Rickettsial, and Chlamydial Infections" (E. H. Lennette and N. J. Schmidt, eds.), pp. 611–632. Am. Public Health Assoc., Washington, DC.
Collier, L. H. (1951). *In* "Freezing and Drying" (R. J. C. Harris, ed.), pp. 133–137. Institute of Biology, London.
Collins, M. K., Patel, P., Flemming, C. L., and Eccles, S. A. (1992). *Bone Marrow Transplant* **9,** Suppl. 1, 171–173.
Culver, K. W., Ram, Z., Wallbridge, S., and Ishii, H. (1992). *Science* **256,** 1550–1552.
Dimmock, N. J. (1967). *Virology* **31,** 338–353.
Dowdle, W. A., Kendal, A. P., and Noble, G. R. (1979). *In* "Diagnostic Procedures for Viral, Rickettsial, and Chlamydial Infections" (E. H. Lennette and N. J. Schmidt, eds.), pp. 585–609. Am. Public Health Assoc., Washington, DC.
Fasquelle, R., and Barbier, P. (1950). *C.R. Seances Soc. Biol. Ses. Fil.* **144,** 1618–1621.
Feinstone, S. M., Barker, L. F., and Purcell, R. H. (1979). *In* "Diagnostic Procedures for Viral, Rickettsial, and Chlamydial Infections" (E. H. Lennette and N. J. Schmidt, eds.). Am. Public Health Assoc., Washington, DC.
Fernie, B., and Gerin, J. (1980). *Virology* **106,** 141–144.
Fox, J. P., Lennette, E. H., Manso, C., and Souza Aguias, J. B. (1942). *Am. J. Hyg.* **36,** 117–142.
Greiff, D. (1991). *In* "Developments in Biological Standardization, Biological Product Freeze Drying and Formulation" (J. May and F. Brown, eds.), pp. 85–92. Karger, Switzerland.
Greiff, D., and Rightsel, W. A. (1967). *Cryobiology* **3**(6), 432–444.
Greiff, D., Rightsel, W. A., and Schuler, E. E. (1964). *Nature (London)* **202,** 624–625.

Grose, C., Friedrichs, W. F., and Smith, K. O., (1981). *Intervirology* **15**(3), 154–160.

Hopps, H. E., and Parkman, P. D. (1979). *In* "Diagnostic Procedures for Viral, Rickettsial, and Chlamydial Infections" (E. H. Lennette and N. J. Schmidt, eds.). Am. Public Health Assoc., Washington, DC.

Horsfall, F. L. (1940). *J. Bacteriol.* **40**, 559–568.

Johnson, H. N. (1979). *In* "Diagnostic Procedures for Viral, Rickettsial, and Chlamydial Infections" (E. H. Lennette and N. J. Schmidt, eds.). Am. Public Health Assoc., Washington, DC.

Kew, O. (1989). *In* "Temperature-Stable Vaccines for Developing Countries" Summary of a Workshop, Institute of Medicine. pp. 125–131. National Academy Press, Washington, DC.

Law, T. J., and Hull, R. N. (1968). *Proc. Soc. Exp. Biol. Med.* **128**, 515–518.

Mauler, R., and Grushkau, H., (1978). *Dev. Biol. Stand.* **41**, 267–70.

Melnick, J. L., Ashkenazi, A., Midulla, V. C., Wallis, C., and Bernstein, A. (1963). *JAMA, J. Am. Med. Assoc.* **185**, 406.

Moss, B., and Flexner, C. (1987). *Annu. Rev. Immunol.* **5**, 305–324.

Penna, H. A. (1956). "Yellow Fever Vaccination." World Health Organization, Geneva.

Rightsel, W. A., and Greiff, D. (1967). *Cryobiology* **3**(6), 423–431.

Rivers, T. M. (1927). *J. Exp. Med.* **45**, 11–21.

Rivers, T. M., and Ward, S. M. (1935). *J. Exp. Med.* **62**, 549–560.

Robin, Y., Saenz, A. C., Outschoorn, A. S., and Grab, B. (1971). *Bull. W.H.O.* **44**, 729–737.

Roemer, K., and Friedmann, T. (1992). *Eur. J. Biochem.* **208**(2), 211–225.

Rombaut, B., Andries, K., and Boeye, A. (1991). *J. Gen. Virol.* **72**, 2153–2157.

Rowe, T. W. G., and Snowman, J. W. (1976). "Edwards Freeze-Drying Handbook." Edwards High Vacuum, Crawley, England.

Scott, E. M., and Woodside, W. (1976). *J. Clin. Microbiol.* **4**, 1–5.

Shackell, L. F. (1909). *Am. J. Physiol.* **24**, 325–340.

Shope, R. E., and Sather, G. E. (1979). *In* "Diagnostic Procedures for Viral, Rickettsial, and Chlamydial Infections" (E. H. Lennette and N. J. Schmidt, eds.), pp. 767–814. Am. Public Health Assoc., Washington, DC.

Simione, F. P., and Brown, E. M., eds. (1991). "ATCC Preservation Methods: Freezing and Freeze-drying," 2nd ed. Am. Type Cult. Collect., Rockville, MD.

Sparkes, J. D., and Fenje, P. (1972). *Bull. W.H.O.* **46**, 729–734.

Turner, T. B. (1938). *J. Exp. Med.* **67**, 61–78.

Wallis, C., and Melnick, J. L. (1968). *J. Virol.* **2**(9), 953–954.

Widdus, R. *et al.* (1989). "Temperature-Stable Vaccines for Developing Countries: Significance and Development Strategies," Summary of a Workshop, Institute of Medicine. pp. 1–9. National Academy Press, Washington, D.C.

CHAPTER 10

Plant Germplasm

H. R. Owen

Background

The manipulation of plant cells, tissues, and organs *in vitro* is producing an increasing number of unique clones of industrial, biochemical, genotypic, and agronomic importance. Examples include regenerable genotypes, transformants, haploids, polyploids, mutants, isogenic lines, somaclonal variants, somatic hybrids, and secondary product-producing cultures.

In addition, a wide array of industrial chemicals is derived from plants, including flavors, pigments, gums, resins, waxes, dyes, essential oils, edible oils, agrochemicals, enzymes, anesthetics, analgesics, stimulants, sedatives, narcotics, and anticancer agents (Wilkes, 1984). Secondary metabolites have been produced from a number of plant species *in vitro,* and many of them have biochemical importance (Fowler and Scragg, 1988). The combination of plant tissue culture and fermentation technology for the production of biochemicals has been reviewed (Zenk, 1978).

Prolonged maintenance of elite plant cells and tissues by repeated subculture is expensive, time consuming, labor intensive, and often results in a reduction in morphogenic or biosynthetic capacity and changes in genetic, chromosomal, or genomic composition, such as mutations, aneuploidy, and polyploidy (D'Amato, 1975). Thus, there is a need to develop and utilize storage methods that reduce the maintenance requirements of plant cultures, while maintaining genetic, biochemical, and phenotypic stability (Benson and Harding, 1990; Withers, 1987b).

Plant preservation initiatives have, by necessity, focused on the conservation of species and land races of international agricultural importance and, to a lesser extent, endangered and threatened species. Conversely, little effort has been made to collect and preserve systematically the increasing number of genotypes being developed with biotechnological applications. At present, these elite cultures, because they are being maintained by individual researchers or laboratories, are in danger of being lost.

The purpose of this chapter is to highlight several methods for maintenance and storage of plant germplasm, with particular emphasis on those techniques most applicable to the preservation of elite cultures used in biotechnology and the plants derived from them.

Methods of Preservation

Seed Preservation

Seed preservation is useful for long-term maintenance of *in vitro*-derived lines if the culture is regenerable and if the characteristics of importance will be maintained after sexual reproduction and conversion to seed (Withers, 1991a). For most plant species, seed preservation is by far the preferred method for germplasm maintenance and storage. In terms of preservation requirements, seed may be classified as desiccation tolerant (orthodox) or desiccation intolerant (recalcitrant or homeohydrous) (Towill and Roos, 1989; Pammenter *et al.,* 1991). Most agricultural and horticultural species produce desiccation-tolerant seed. In general terms, desiccation-tolerant seeds stored at 5°C have longevities in the range of 5–20 years, whereas those stored at −10 to −20°C have longevities of 20–50 years or more (Towill and Roos, 1989).

Seed viability during storage is influenced by seed harvesting factors, seed moisture content, storage temperature, and genotype. Generally, orthodox seeds are dried to 5–7% moisture content, sealed in moisture-proof containers, and stored between 5 and −20°C (Towill and Roos, 1989). Harrington's rule (Harrington, 1963) states that for each percent decrease in seed moisture content (between 5 and 14%), a seed's life span is doubled. Independently, for each 5°C decrease in temperature (between 0 and 50°C), a seed's life span is doubled. These guidelines, however, are very general and are affected by a number of factors, including genotype, seed dormancy, and homogeneity of seed lots (Ellis, 1984b).

Many orthodox seeds can be dried to low moisture levels and stored in liquid nitrogen (LN) with very high survival rates (Sakai and Noshiro, 1975; Standwood, 1980; Styles *et al.,* 1982; Pence, 1991b). This, however, depends on seed coat properties, seed size, and seed oil content (Standwood, 1985; Towill, 1989; Vertucci, 1989, 1992). Freeze-drying has been used to a limited extent to reduce moisture contents and improve storability for some seed (Woodstock *et al.,* 1983). In addition, alternative methods to extend seed storage by modifying the gaseous environment have been reported (Syroedov *et al.,* 1987; Senft, 1989; Sowa *et al.,* 1991; Sowa and Towill, 1991a).

Several tests have been employed to estimate seed vigor after storage, including accelerated aging, artificial aging, controlled deterioration, the cold test, the conductivity test, seedling growth rate, and tetrazolium testing (Towill and Roos, 1989; Roos and Wiesner, 1991). The entire subject of orthodox seed storage has been extensively reviewed (Roberts, 1975; Stanwood and Bass, 1981; Ellis, 1984a) and a detailed discussion of seed storage is outside the scope of this chapter. For specific recommendations on seed storage procedures and facilities, the reader is directed to several excellent manuals [Justice and Bass, 1978; Stanwood and Bass, 1978; Cromarty *et al.,* 1985; Ellis *et al.,* 1985; Hanson, 1985; International Board for Plant Genetic Resources (IBPGR), 1985].

Vegetative Propagation

A number of plant species and selections cannot be maintained as botanical seed. Many tropical species produce desiccation-intolerant seed and may have life spans of only weeks to months (Farrant *et al.,* 1988; Towill and Roos, 1989). A significant number of economically important species produce recalcitrant seed, including avocado, cocoa, coconut, rubber, mango, wild rice, tea, sugar maple, palm, chestnut, sycamore, cinnamon, nutmeg, mahogany, and oak (King and Roberts, 1980a,b; Ellis, 1984a; Towill and Roos, 1989; Williams, 1989; Shell, 1990; Withers *et al.,* 1990). Germplasm also must be maintained vegetatively for species and selections in which maintenance of heterozygosity is essential, for species that do not set seed easily or set enough seed, for species with long juvenile periods, and for sterile lines (Towill and Roos, 1989). These types of plant material would benefit from *in vitro* maintenance and preservation methods. A number of chapters and reviews are suggested as additional information sources (Chin and Roberts, 1980; King and Roberts, 1980a,b; Roberts and King, 1980; Roberts *et al.,* 1984; Towill, 1988; Chin, 1988; Roos, 1989; Ford-Lloyd, 1990).

In Vitro Maintenance and Storage

GROWTH REDUCTION

The growth of plant cells and tissues *in vitro* can be reduced by a number of methods, thus increasing the subculture interval. The principle behind reduced-growth maintenance of *in vitro* cultures is based on the modification of the culture environment to reduce growth rates while maintaining viability (Ng and Ng, 1991). Thus, the costs and potential problems encountered during subculturing are reduced. Several strategies are noted below. It is important to stress that a particular treatment, or combination of treatments, may or may not be tolerated by a particular species or culture and should be tested in each case.

Physical Environment The most common method used to suppress cell division and DNA synthesis (and possibly reducing water loss) is by culturing at reduced temperatures, usually at or near the minimum critical temperature for a particular species (Kartha, 1985). Deep undercooling of cell cultures in an oil emulsion to subfreezing temperatures without ice formation has also been examined (Mathias *et al.*, 1985). Other manipulations of the incubation temperature that were shown to be effective in certain cases include stepwise reduction in temperature to acclimate cultures to reduced temperatures, and periodic warming of cultures during the storage period (Staritsky *et al.*, 1986).

A reduction in oxygen content or atmospheric pressure has been used to reduce growth rates (Bridgen and Staby, 1981). Mineral oil (Caplin, 1959; Augereau *et al.*, 1985), paraffin oil (Augereau *et al.*, 1986), and silicone oil overlays (Moriguchi *et al.*, 1988) have been effective in reducing callus growth rates, either by reducing respiration rates or by decreasing water losses. Containers that allow gaseous exchange, but minimize water loss, also have extended subculture intervals (Reed, 1991, 1992). Other modifications of the physical environment that have been examined include callus desiccation (Nitzsche, 1978), reduced light (Kartha, 1981; Schaper and Zimmer, 1989), and dark culture (Ng and Ng, 1991).

Culture Medium Growth also can be slowed by the addition of osmotically active substances, such as elevated sucrose or agar levels, or by the addition of mannitol (Wanas and Callow, 1986) or sorbitol (Gunning and Lagerstedt, 1986) to the culture medium. Growth regulators and inhibitors, such as 2,2-dimethylhydrazine, 2-chloroethyl trimethylammonium chloride (Gunning and Lagerstedt, 1986), *N*-dimethylsuccinamic acid (Mix, 1982), maleic hydrazide, diaminozide, chlorocholine chloride, and abscisic acid (Lizarraga *et al.*, 1989) have also been used.

Other manipulations to extend subculture intervals include periodic addition of liquid medium (Mullin and Schlegel, 1976), serial microtuber induction (Kwiatkowski *et al.*, 1988), reduced nitrate (Moriguchi and Yamaki, 1989) or cytokinin (Bertrand-Desbrunais *et al.*, 1991) concentrations, reduced basal medium concentration, increased culture medium volume relative to explant size, and substitution of distilled water for the culture medium (Zee and Munekata, 1992).

Applications   The advantages of *in vitro* preservation include maintenance of disease-free lines, rapid clonal multiplication rates, smaller storage
space requirements, continuous availability, and ease of shipment (Towill,
1988; Towill and Roos, 1989). The disadvantages of *in vitro* maintenance
are relatively high inputs of time and labor for culture establishment and
maintenance, potential losses due to contamination or mislabeling, genotypic variability, loss of regenerative capacity, and culture-induced (somaclonal) genetic variation (Towill, 1988; Towill and Roos, 1989).

Cost comparisons between *in vitro* maintenance and conventional *in
planta* maintenance are few (Jarret and Florkowski, 1990). *In vitro* culture
requires strict control of environmental conditions, such as medium constituents, and often such conditions are not immediately applicable to a wide
range of species or even selections within a species (Reed, 1990a). Thus,
culture conditions often have to be developed for each particular species and
culture of interest. The International Board for Plant Genetic Resources
(IBPGR) has published general recommendations for *in vitro* storage, including recommendations on the design and operation of culture facilities
(IBPGR, 1986).

Genetic Stability   Prolonged maintenance of dedifferentiated plant
cells and tissues *in vitro* by repeated subculture often results in somaclonal
variation, and it may manifest itself at the molecular, biochemical, or phenotypic level (Benson and Harding, 1990). Somaclonal variation can be modified to a limited extent by environmental factors, such as medium formulation and subculture interval (Reisch, 1988), but cannot be eliminated with
certainty unless metabolism is suspended (Withers, 1990). The type of
explant and the preservation method can have a significant impact on
survival and the extent of somaclonal variation. In general, organized cultures (meristems, shoot tips, and embryos) are more stable than unorganized cultures (protoplasts, suspensions, and calli). Thus, organized cultures
have a better likelihood of retaining their genetic integrity during prolonged
culture *in vitro* (Snowcroft, 1984). For short to medium-term maintenance,
cell and tissue cultures may benefit from some form of growth reduction,
but for long-term maintenance, growth suspension must be recommended
(Hiraoka and Kodama, 1984).

GROWTH SUSPENSION

Growth suspension by cryopreservation [termed *viva-cryopreservation*
by Finkle *et al.* (1985b)] is recommended as the most effective way of
preserving *in vitro* cultures for extended periods (Withers, 1985b; Kartha,
1987; Benson and Harding, 1990). Cryogenic storage refers to storage below
$-130°C$, where liquid water is absent and molecular kinetics and diffusion
rates are extremely low (James, 1983; Towill and Roos, 1989). In practice,

this refers to storage in or over liquid nitrogen ($-196°C$ liquid phase, $-150°C$ vapor phase). Advantages of cryopreservation include indefinite storage without subculturing, frequent viability testing, and plant regeneration (Towill, 1988) and maintenance of biosynthetic and regeneration capacity of cultures over time (Shillito et al., 1989). It would also facilitate the continued utilization of unique lines, allow for more highly controlled time course studies, and avoid the costly reinitiation of desirable lines. The only potential disadvantage of cryopreservation is the need to maintain a low-temperture environment without interruption (DiMaio and Shillito, 1989).

The list of species and explant types that have been successfully cryopreserved has increased substantially in the past 5–10 years and it now includes more than 100 species. It is also encouraging that postthaw viability percentages are increasing. Due to the enormous variability that exists within the plant kingdom, however, it is unlikely that a single procedure will be identified that is applicable without modification to a large number of species. An exception may be in the advances realized in the cryopreservation of homogeneous suspension cultures, where somewhat standardized protocols have been detailed (Withers, 1991b). Many of the protocols for plant cryopreservation have been developed by an empirical approach. With a better understanding of freezing tolerance, cryoprotectant action, and freezing damage, it may be possible to develop a limited number of cryopreservation protocols that would be applicable to a wider range of plant species.

Selected cryopreservation methods by plant species and explant type are listed in Table 1 and are provided as suggested starting points for further investigations. The following sections are not intended to cover the subject of cryopreservation in great detail, but rather to discuss briefly the steps involved. For a more exhaustive discussion of the topic and for model protocols, the readers are directed to several comprehensive papers (Kartha, 1981; Withers, 1985a, 1987a, 1990; Chen and Li, 1989) and books (Kartha, 1985; Grout and Morris, 1987; Li, 1989).

Starting Material The physiological state and genotype of the material to be preserved determine to a great extent the success or failure of cryopreservation (Kartha, 1981). *In vitro* cultures display a wide heterogeneity in cell synchrony and morphology (Kartha, 1987). Recently established cultures that are in the late log phase or early division phase of growth contain a larger population of small, densely cytoplasmic cells and, therefore, are better able to withstand cryoprotection and freezing (Bajaj, 1976a; Withers and Street, 1977). Similarly, meristematic tissues (shoot meristems) contain a population of dense, actively dividing cells. Organized cultures (meristems, shoot tips) have been shown to be more difficult to cryopreserve compared to cells or callus tissues, but this probably relates to difficulties in cryoprotectant penetration or uniformity of freezing than to cell type. As a result of incomplete survival, cryopreserved shoot tips may form callus or adventitious organs (shoots, roots, or somatic embryos), and thus would have an increased likelihood of genetic instability (Withers *et al.,* 1990). In general, meristems isolated from temperate species are more tolerant to freezing than are tropical species (Kartha, 1981).

Preconditioning Preconditioning of cultures for various times at reduced temperatures, or elevated osmotic pressures has been shown to increase freezing resistance in some systems (Kartha, 1987). This is usually accomplished by culturing material at temperatures near the chilling sensitivity temperature for a particular species, or by incubating cultures in a medium containing osmotically active compounds such as sucrose, mannitol, sorbitol, or proline (Withers and King, 1980; Kartha, 1987).

Cryoprotectant Treatment With the exception of shoots of some naturally cold-hardened species or artificially cold-hardened cultures (Sakai and Sugawara, 1973), the freeze-preservation of vegetative plant material requires use of cryoprotectants to minimize the occurrence of lethal intracellular ice formation (Finkle *et al.,* 1985b). A number of compounds have been shown to be effective, including dimethyl sulfoxide (DMSO), glycerol, sucrose, glucose, trehalose, sugar alcohols, diols, polyols, and amino acids (Kartha, 1987; de Boucaud and Cambecedes, 1988; Bakas and Disalvo, 1991).

Toxicity and genetic damage may be a problem due to the required high concentration of cryoprotectants (Ashwood-Smith, 1985; Towill and Roos, 1989), and chemical impurities have been shown to contribute to toxic effects (Matthes and Hackensellner, 1981). In many studies, however, a synergistic effect has been observed when combinations of cryoprotectants are used, and this may allow the concentrations of certain components to be reduced to below individual toxic levels without reducing overall efficacy (Benson and Harding, 1990).

Vitrification (glass formation without ice crystallization) can be achieved with certain cryoprotectant solutions at high concentration and by very rapid cooling rates, and this may confer protection during freezing in some systems by stabilizing the organization of water (James, 1983; Fahy *et al.,* 1987; Towill and Roos, 1989).

Generally, DMSO, either alone or in combination with other protectants, has been effective in conferring freeze-protection to organized cultures (meristems and shoot tips). Preservation of unorganized cultures, such as cell suspensions and callus cultures, generally requires a combination of protectants (Withers, 1987a,b). In some cases, cryoprotectants dissolved in culture medium have been more effective than solutions prepared in distilled water (Withers, 1980). Generally, a slow or stepwise addition of cryoprotectants is performed to minimize osmotic shock.

TABLE 1

Selected Cryopreservation Methodologies by Plant Species and Explant Type

Species	Explant[a]	Pretreatment and protectant[b]	Freezing[c]	Recovery	Regeneration[d]	Reference
Acer pseudoplatanus	Su	0.5 *M* DMSO, 0.5 *M* glycerol, 1 *M* Pro	−1°C/min to −35°C, then LN	55–60%	NR	Pritchard et al. (1986)
Aesculus sp.	ST	Desiccate in air overnight	LN immersion	Yes	Some	Pence (1990)
Aesculus hippocastanum	ZE	Desiccation	LN immersion	Up to 80%	NR	Pence (1988)
	EA	Desiccation	LN immersion	Up to 100%	Yes	Pence (1992)
Arabidopsis thaliana	Pr	1 *M* Suc, 0.5 *M* glycerol, 0.5 *M* DMSO	−1°C/min to −60°C, then LN	Yes	Yes	Ford (1990)
Arachis hypogaea	PE	Preculture 4–6 weeks, then 5% DMSO, 5% Suc, 5% glycerol	LN immersion	29%	NR	Bajaj (1983c)
Arachis villosa	ST	5% Suc, 5% glycerol, 5% DMSO	LN immersion	23–31%	Yes	Bajaj (1979)
	PE	Preculture 4–6 weeks, then 5% DMSO, 5% Suc, 5% glycerol	LN immersion	38%	Yes	Bajaj (1983c)
Araucaria hunsteinii	ZE	Desiccate in air to 20% H_2O	LN immersion	80%	2°	Pritchard and Prendergast (1986)
Asparagus officinalis	SE	Preculture 0.3 *M* Sor, 0.2 *M* Suc 16 hours, then 12% EG, 0.5 *M* Sor	LN immersion	45–48%	Yes	Uragami et al. (1989)
	ST	Preculture 0.7 *M* Suc 2 days, then desiccate to 20% H_2O	LN immersion	63%	Yes	Uragami et al. (1990)
Atropa belladonna	Su	0.4 *M* Suc, 2 *M* glycerol	−30°C 1 hour, then LN	78%	Yes	Nishizawa et al. (1992)
	PE	7% DMSO or 15% glycerol	−2°C/min, then LN	33%	Yes	Bajaj (1978)
	Pr	10% DMSO, 1 *M* Man, 4% Suc	Gradual LN immersion	Yes	Yes	Bajaj (1988)
	Su	Preculture 5% DMSO, then 5% DMSO	−2°C/min to −100°C, then LN	30–40%	NR	Nag and Street (1975a,b)
Beta vulgaris	Me	Preculture 5% DMSO, then DMSO-Sor	To 0°C then −0.5–1°C/min	Up to 58%	NA	Braun (1988)
Betula sp.	ST	Winter twigs precooled to −5°C	−5°C/hour stepwise to −15°C, then LN	Yes	NA	Sakai (1960, 1965)
Brassica campestris	PE	Preculture 4–6 weeks, then 5% DMSO, 5% Suc, 5% glycerol	LN immersion	31%	Yes	Bajaj (1983c)
Brassica napus	Su	1.5M EG, 0°C	LN immersion	40%	NR	Langis et al. (1989)
	Mi	13% Suc	−15°C/hour to −15°C, then LN	Yes	Yes	Charne et al. (1988)
	PE	Preculture 4–6 weeks, then 5% DMSO, 5% Suc, 5% glycerol	LN immersion	44%	NR	Bajaj (1983c)
	ST	Preculture 5% DMSO 48 hours, then 15% DMSO	−1°C/min to 0°C, then LN	24%	NA	Benson and Noronha-Dutra (1988)
	ST	Preculture 5% DMSO 24 hours, then 15% DMSO	LN immersion	Yes	Yes	Withers et al. (1988)
	Su	Preculture 1 *M* Sor	−1°C/min to −35°C, then LN	Yes	NR	Weber et al. (1983)

Species	Type	Cryoprotectant/Treatment	Freezing protocol	Survival	Regrowth	Reference
Brassica oleracea var. *gemmifera*	ST	1.5 M DMSO, 3% Glc, 20°C 2 hours	−0.5°C/min to −10°C, nucleate, −0.5°C/min to −30°C, then LN	85%	Yes	Harada et al. (1985)
Bromus inermis	Pr, Su	0.7 M glycerol	Stepwise 4°C, −20°C, −60°C, LN	Yes	NR	Mazur and Hartmann (1979)
Camellia sinensis	EA	Desiccate to 10–13% H₂O	LN immersion	83–95%	Yes	Chaudhury et al. (1991)
Cannabis sativa	Su	10% DMSO	−2°C/min to −10°C, then LN	58%	NR	Jekkel et al. (1989)
Carya sp.	ST	Desiccate in air overnight	LN immersion	Yes	Some	Pence (1990)
	ZE	Desiccation	LN immersion	Up to 100%	NR	Pence (1988)
Castanea sp.	ST	Desiccate in air overnight	LN immersion	Yes	Some	Pence (1990)
	ZE	Desiccation	LN immersion	Up to 80%	NR	Pence (1988)
Catharanthus roseus	Pr	0.55 M Sor, 0.7M DMSO	−1°C/min to −40°C, then LN	20%	NR	Gazeau et al. (1992)
	Su	Preculture 1 M Sor 24 hours, then 5% DMSO	−0.5C/min to −40°C, then LN	Yes	NR	Mannonen et al. (1990)
Chamomilla recutita	ST	0.5 M DMSO, 1 M glycerol, 0.2 M Suc	−0.5°/min to −40°C, then LN, or LN immersion	65%	40%	Diettrich et al. (1990)
Cicer arietinum	ST	5% Suc, 5% glycerol, 5% DMSO	LN immersion	27–36%	Yes	Bajaj (1979)
Chicorium intybus	ST	Preculture 2 days, then 15% DMSO 1 hour	−0.5°C/min to −40°C, then LN	39–50%	NA	Demeulemeester et al. (1993)
Citrus sp.	Ov	7% DMSO, 7% Suc	LN immersion	29%	Yes	Bajaj (1984)
Citrus sinensis	Su	30% glycerol, 15% EG, 15% DMSO	LN immersison	80%	Yes	Sakai et al. (1990)
Cocos nucifera	ZE	Transverse halves 7% DMSO, 7% Suc	LN immersion	25%	NR	Bajaj (1984)
	ZE	Desiccate in air, then in 600 g/liter Suc, 15% glycerol	LN immersion	33–93%	Yes	Assy-Bah and Engelmann (1992)
Coffea arabica	SE	Preculture 0.75 M Suc, then 0.75 M Suc, 5% DMSO stepwise	−0.5°C/min to −11°C, nucleate, then to −40°C, then LN	45–50%	2°	Bertrand-Desbrunais et al. (1988)
Coffea canephora	ZE	Desiccate in air	LN immersion	41%	Yes	Abdelnour-Esquivel et al. (1992a)
Coffea liberica	ZE	Desiccate in air	LN immersion	85%	Yes	Normah and Vengadasalam (1992)
Coleus blumei	Su	1 M Sor 8 hours	−1–2°C min⁻¹	35%	NR	Reuff et al. (1988)
Corylus sp.	ST	Desiccate in air overnight	LN immersion	Yes	Some	Pence (1990)
Cucumis melo	SE	Preculture ABA 3 days, desiccate to 50–65% H₂O	LN immersion	14–65%	Yes	Shimonishi et al. (1991)
Datura innoxia	Pr	10% DMSO, 1 M Man, 4% Suc	Gradual LN immersion	Yes	Yes	Bajaj (1980)
	Su	Preculture 1 M Sor	−1°C min⁻¹ to −35°C, then LN	Yes	NR	Weber et al. (1983)
Datura stramonium	Su	7% DMSO	−1°C min⁻¹ to −100°C, then LN	Yes	NR	Bajaj (1976b)
Daucus carota	Pr	10% DMSO, 10% glycerol	0°C, −1°C/min to −40°C, hold 1 hour, then LN	Yes	NR	Hauptmann and Widholm (1982)
	SE	0.4 M Suc	−20°C for 24 hours, then LN	80%	43%	Lecouteux et al. (1991)
	Su	0.5 M DMSO, 0.5 M glycerol, 1 M Suc	−1°C/min to −35°C 40 min, then LN	75–80%	NR	Benson and Noronha-Dutra (1988)

(Continues)

TABLE 1 (*Continued*)

Species	Explant[a]	Pretreatment and protectant[b]	Freezing[c]	Recovery	Regeneration[d]	Reference
Dendrathema grandiflorum	ST	Preculture 2% Suc, 5% DMSO 2 days, then 10% DMSO, 3% Glc	$-0.2°C/min$ to $-40°C$, then LN	87%	47%	Fukai (1990); Fukai *et al.* (1991a)
Dianthus sp.	ST	10% DMSO, 3% Glc	$-0.5°C/min$ to $-40°C$, then LN	92–100%	46–100%	Fukai *et al.* (1991b)
Dianthus caryophyllus	Me	Preculture 0.5 *M* Suc, then 0.5 *M* Suc, 5% DMSO 2 hours, 0°C	$-0.5°C/min$, the LN	79–94%	NA	Dereuddre *et al.* (1987)
	ST	Preculture 0.75 *M* Suc, then 0.75 *M* Suc, 5–15% DMSO	$-0.5°C/min$ to $-40°C$, then LN	Nearly 100%	NA	Dereuddre *et al.* (1988)
Digitalis lanata	ST	Preculture 4°C 8 weeks, then 2*M* DMSO	$-0.5°C/min$ to $-40°C$, then LN, or LN immersion	70%	30%	Diettrich *et al.* (1987)
	Su	Preculture 6% Man 3 days, then 0.5 *M* DMSO, 0.5 *M* glycerol, 1 *M* Suc	$-1°C/min$ to $-35°C$, then LN	>50%	NR	Seitz *et al.* (1983)
Dioscorea deltoidea	Su	Preculture Suc or amino acids, various protectants	To $-5–7°C$, nucleate, $-0.5°C/min$ to $-30°C$, $-9°C/min$ to $-70°C$, then LN	Yes	NR	Butenko *et al.* (1984)
Distichlis sp.	Su, Pr	Glc–Suc–Fru or 0.55 *M* Man	To $-2°C$, nucleated, $-5°C/hour$	Yes	NR	Bartolo *et al.* (1987)
Elaesis guineenis	SE	0.75 *M* Suc 7 days	LN immersion	Yes	Yes	Engelmann *et al.* (1985); Engelmann and Dereuddre (1988); Engelmann (1990a,b)
Fagus sp.	ZE	Desication to 10.4% H$_2$O	LN immersion	100%	Yes	Grout *et al.* (1983)
	ST	Desiccate in air overnight	LN immersion	Yes	Some	Pence (1990)
	ZE	Desication	LN immersion	Up to 100%	NR	Pence (1988)
Fagus sylvatica	PE	2% DMSO, 2.5% Suc	0°C, then $-10°C$, then $-0.5°C/min$ to $-40°C$, then LN	90%	Yes	Jörgensen (1991)
Fragaria × *ananassa*	Me	Preculture 15% DMSO 2 days	$-0.84°C/min$ to $-40°C$, then LN	55–95%	Yes	Kartha (1980)
Glycine max	Ca	Preculture 1.25 *M* Suc	$-0.5°C/min$ to $-40°C$, then LN	Yes	NR	Engelmann (1992)
	Pr	0.7 *M* Sor	$-1°C/min$ to $-35°C$, then LN	Yes	NR	Weber *et al.* (1983)
	Su	5% DMSO	$-2°C/min$ to $-100°C$, then LN	50%	NR	Bajaj (1976b)
Gossypium arboreum	An, Ov	3% DMSO 3 days, then 5% DMSO, 5% Suc, 5% glycerol	LN immersion	34–42%	NR	Bajaj (1982)
Gossypium hirsutum	An, Ov	3% DMSO 3 days, then 5% DMSO, 5% Suc, 5% glycerol	LN immersion	Yes	NR	Bajaj (1982)
Grevillea scapigera	ST	Preculture 5% DMSO, the 10% DMSO	$-0.5°C/min$ to $-40°C$, then LN	20%	Yes	Touchell *et al.* (1992)
Gypsophila sp.	ST	10% DMSO, 3% Suc	$-0.5°C/min$ to $-40°C$, then LN	90–100%	41–97%	Fukai *et al.* (1991b)

Species	Type	Cryoprotectant/pretreatment	Freezing protocol	Survival	Regrowth	Reference
Haplopappus gracilis	ST	Preculture 5% DMSO 3 days, then 10% DMSO stepwise	−0.5°C/min to −40°C, then LN	Yes	Yes	Taniguchi *et al.* (1988)
Hevea brasiliensis	ZE	Desiccate to 16% H₂O	LN immersion	Yes	87%	69%	Normah *et al.* (1986)
Hordeum vulgare	Ca	0.5 *M* DMSO, 0.5 *M* glycerol, 1 *M* Suc 4°C 2 hours, then blot	−3°C/min to −80°C, then LN	78%	Yes	Hahne and Lorz (1987)
Howea fosteriana	ZE	Desication to 10% H₂O	LN immersion	63–64%	Yes	Chin *et al.* (1988)
Hyoscyamus muticus	Su	0.5 *M* glycerol, 0.5 *M* DMSO, 1 *M* Suc	−1°C/min to −35°C, then LN	Yes	NR	Withers and King (1980)
Ipomoea batatas	ST	30% glycerol, 15% EG, 15% DMSO	LN immersion	Yes	Yes	Towill and Jarret (1992)
Juglans sp.	ST	Desiccate in air overnight	LN immersion	Yes	Some	Pence (1990)
Juglans sp.	ZE	Desiccation	LN immersion	Up to 100%	NR	Pence (1988)
Larix × eurolepis	Su	0.4 *M* Sor 24 hours, then 10% DMSO	−0.33°C/min to −40°C, then LN	Yes	Yes	Klimaszewska *et al.* (1992)
Lavandula vera	Ca	10% Glc, 5% DMSO	−1°C/min to −40°C, then LN	Yes	NR	Kuriyama *et al.* (1990)
Luchnis sp.	ST	10% DMSO, 3% Glc	−0.5°C/min to −40°C, then LN	81–100%	21–84%	Fukai *et al.* (1991b)
Lycopersicon esculentum	Me	15% DMSO	Stepwise LN immersion	45%	2°	Grout *et al.* (1978)
Malus domestica	ST	Dormant buds prefrozen to −15°C	LN immersion	Yes	NR	Katano *et al.* (1983)
Malus domestica	ST	Desiccate to 20–30% H₂O at −4°C	−2°C/hour to −16°C, −10°C/hour to −30°C 24 hours, then LN	Yes	Yes	Tyler and Stushnoff (1988a,b)
Manihot esculenta	Me	Preculture 3% DMSO 5–7 days, then 5% DMSO, 5% Suc, 5% glycerol	LN immersion	29%	Yes	Bajaj (1983a)
Manihot utilissima	ZE	None	LN immersion, slow stepwise thaw	97%	34%	Marin *et al.* (1990)
Manihot utilissima	Me	5% Suc, 10% glycerol	LN immersion	21%	13%	Bajaj (1977)
Marchantia polymorpha	Pr	Preculture 20 hours 0.23 *M* Man	−0.5–0.8°C/min	40%	NR	Sugawara and Tekeuchi (1988)
Medicago sativa	Ca	10% PEG, 8% Glc, 10% DMSO	−1°C/min to −30°C, then LN	Yes	Yes	Finkle *et al.* (1985a)
Mentha sp.	ST	Stepwise 35% EG, 1 *M* DMSO, 10% PEG 8000	LN immersion	56%	44%	Towill (1990)
Morus sp.	ST	Preculture 5°C 20–40 days, then 10% DMSO, 0.5 *M* Sor	−1°C/min to −42°C, then LN	Yes	Yes	Niino and Oka (1990)
Morus Bombycis	ST	Intact buds	Stepwise freezing (−10°C/day) to −10°C, hold 6 hours, then LN	50%	Yes	Yakuwa and Oka (1988)
Musa sp.	Su	7.5% DMSO	−1°C/min to −10°C, nucleate, then to −40°C, then LN	92%	Yes	Panis *et al.* (1990)
Musa acuminata	ZE	Desiccate in air	LN immersion	83%	Yes	Abdelnour-Esquivel *et al.* (1992b)
Musa balbisiana	ZE	Desiccate in air	LN immersion	92%	Yes	Abdelnour-Esquivel *et al.* (1992b)
Nicotiana plumbaginifolia	Su	Preculture 5–10% trehalose, then 40% trehalose	−1°C/min to −40°C, then LN	Yes	NR	Bhandal *et al.* (1985)
Nicotiana sylvestris	Su	Preculture 6% Sor 2–5 days, then 0.5 *M* DMSO, 0.5 *M* glycerol	Slow to −40°C, then LN	75%	NR	Maddox *et al.* (1982-1983)

(Continues)

TABLE 1 (*Continued*)

Species	Explant[a]	Pretreatment and protectant[b]	Freezing[c]	Recovery	Regeneration[d]	Reference
Nicotiana tabacum	PE	7% DMSO or 15% glycerol	−2°C/min, then LN	33%	Yes	Bajaj (1978)
	Pr	10% DMSO, 1 *M* Man, 4% Suc	Gradual LN immersion	Yes	Yes	Bajaj (1988)
	Su	Preculture 5–10% trehalose, then 40% trehalose	−1°C/min to −40°C, then LN	Yes	NR	Bhandal *et al.* (1985)
Oryza sativa	An	5% DMSO, 5% Suc, 5% glycerol	LN immersion	Yes	Yes	Bajaj (1980)
	Ca	10% Glc, 5% DMSO	−1°C/min to −30°C, then LN	Yes	NR	Kuriyama *et al.* (1990)
	Pr, Su	5% DMSO	Slow to −70°C, then LN	Yes	NR	Cella *et al.* (1982)
	Su	10% PEG, 8% Glc, 10% DMSO	−1°C/min to −23 or −30°C, then LN	Yes	NR	Finkle and Ulrich (1982)
	Su	Preculture 0.33 *M* Man 3 days, then 1 *M* Suc, 0.4 *M* glycerol, 0.045 *M* Pro	−1°C/min to −35°C, then LN	Yes	Yes	Meijer *et al.* (1991)
Oryza sativa × Pisum sativum	Pr	0.5 *M* Man, 5% DMSO	LN vapors	12%	NR	Bajaj (1983d)
Panax ginseng	Su	Preculture Suc or amino acids, various protectants	To −5–7°C, nucleate, −0.5°C/min to −30°C, −9°C/min to −70°C, then LN	Yes	NR	Butenko *et al.* (1984)
	Su	Preculture 6% Man 3 days, then 0.5 *M* DMSO, 0.5 *M* glycerol, 1 *M* Suc	−1°C/min to −35°C, then LN	Yes	NR	Mannonen *et al.* (1990)
Panicum maximum	Su	Preculture 0.33 *M* Man 3 days, then 0.5 *M* Sor, 0.7 *M* DMSO	−0.5°C/min to −40°C, then LN	99%	NR	Gnanapragasam and Vasil (1992)
Papaver somniferum	Su	0.5 *M* Sor, 5% DMSO	−0.5°C/min to −35°C, then LN	Yes	NR	Friesen *et al.* (1991)
Petunia hybrida	PE	7% DMSO or 15% glycerol	−2°C/min, then LN	33%	Yes	Bajaj (1978)
Phaseolus vulgaris	EA	10% PEG, 8% Glc, 10% DMSO stepwise	LN immersion	59%	NR	Zavala and Sussex (1986)
Phoenix dactylifera	Ca	10% PEG, 8% Glc, 10% DMSO	−3° C/min to −4°C, nucleate, −1°C/min to −30°C, then LN	Yes	Yes	Ulrich *et al.* (1982)
	ST	Preculture 0.1 *M* Suc 1 day, then 0.5 *M* Suc 1 day, then 10% DMSO, 0.5 *M* Suc	−1°C/min to −30°C, then LN	Yes	Yes	Bagniol *et al.* (1992)
Picea glauca	EC	5% DMSO, 0.4 *M* Sor, 2% Suc	−0.5°C/min to −40°C, then LN	Yes	18–35%	Toivonen and Kartha (1989)
	Su	Preculture 0.4 *M* Sor, then 0.4 *M* Sor, 5% DMSO	−0.3°C/min to −35°C, then LN	Yes	Yes	Kartha *et al.* (1988)
Pinus caribaea	Su	Preculture 0.4 *M* Suc, then 5% DMSO	0.C, then −0.5°C/min to −35°C, then LN	Up to 100%	Yes	Lainé *et al.* (1992)
Pinus pumira	ST	Winter twigs precooled to −5°C	−5°C/hour stepwise to −30°C, then LN	Yes	NA	Sakai (1960, 1965)
Pinus sylvestris	ST	None	−1°C/min to −39°C, then LN	Yes	NR	Kuoska and Hohtola (1991)

Pisum sativum	EA	10% glycerol, 1 M Suc, air dry 1 hour	LN immersion	Yes	Yes	Mycock et al. (1991)
	Me	5% DMSO	−0.6°C/min to −40°C, then LN	Yes	Yes	Haskins and Kartha (1980)
Poncirus trifoliata	EA	Desication to 14% H_2O	LN immersion	76%	Yes	Radhamani and Chandel (1992)
Populus sp.	ST	Winter twigs precooled to −5°C	−5°C/hour stepwise to −15 to −20°C, then LN, slow thaw	Yes	NA	Sakai (1960, 1965)
Populus euramericana	Su	5% DMSO, 5% glycerol	−1°C/min to −40°C, then LN	Yes	NR	Binder and Zaerr (1980b)
	Ca	Preculture 15°C (day), 0°C (night) 60 days, then 0°C 20 days	−5°C/day to −20 or −30°C, then LN	Yes	NR	Sakai and Sugawara (1973)
Porphyra yezoensis	LS	None	Prefreeze to −10 or −20°C, then LN	Yes	NA	Sakai and Otsuka (1972)
Primula obconica	An	Preculture 3 weeks, then 7% DMSO, 7% Suc	LN immersion	Yes	Yes	Bajaj (1981b)
Pseudotsuga menziesii	Su	5% DMSO 1% glycerol	−1°C/min to −40°C, then LN	65%	NR	Binder and Zaerr (1980a)
Pyrus sp.	Me	Harden at −1°C (nights) 1 week, then 5% DMSO 48 hours, then 10% PEG, 10% Glc, 10% DMSO	−0.1°C/min to −40°C, then LN	5–95%	Yes	Reed (1990b)
Pyrus communis	Pr, Su	Glc-Suc-Fru or 0.55 M Man	To −2°C, nucleated −5°C/hour	Yes	NR	Bartolo et al. (1987)
	ST	Pretreat 0°C 8 weeks, preculture, 0.75 M Suc, then 0.75 M Suc, 15% DMSO	−0.5°C/min to −30°C, then LN	71%	NA	Dereuddre et al. (1990a,b)
Pyrus pyrifolia	ST	Dormant shoots pretreated −3°C 14 days, −5°C 3 days −10°C 1 day	LN immersion	78%	NA	Sakai and Nishiyama (1978)
Pyrus serotina	ST	None	Stepwise freezing, then LN	80%	20–50%	Oka et al. (1991)
Quercus sp.	ST	Desiccate in air overnight	LN immersion	Yes	Some	Pence (1990)
Quercus faginea	EA	Desiccate in air	LN immersion	60%	Yes	Gonzalez-Benito and Perez-Ruiz (1992)
Quercus petraea	PE	2% DMSO, 2.5% Suc	0°C, then −10°C, then −0.5°C/min to −40°C, then LN	90%	Yes	Jörgensen (1991)
Raphanus sp.	ST	Preculture 4% DMSO 2 days, then 8% DMSO, 8% Suc, or 12% DMSO	−0.25°C/min to −35°C, then LN	50%	NA	Towill (1991)
Ribes grossularia	ST	Preculture 4% DMSO 2 days, then 8% DMSO, 8% Suc, or 12% DMSO	−0.25°C/min to −35°C, then LN	67%	NA	Towill (1991)
Robinia pseudoacacia	Pr	Winter bark cells in 1M Suc	−2°C hour to −4°C, nucleate, then to −40°C, then −10°C/hour to −70°C, then LN	60%	NR	Siminovitch (1979)
Rosa 'Paul's Scarlet'	Su	Preculture 6% Man 4 days, then 1 M DMSO, 2 M Suc, 1 M glycerol	−1°C/min to −35°C, then LN	Yes	NR	Strauss et al. (1985)
Rubus sp.	Me	Harden at −1°C nights 1 week, then 5% DMSO 48 hours, then 10% PEG, 10% Glc, 10% DMSO	−0.8°C/min to −40°C, then LN	51–67%	Yes	Reed (1988)

(Continues)

TABLE 1 (*Continued*)

Species	Explant[a]	Pretreatment and protectant[b]	Freezing[c]	Recovery	Regeneration[d]	Reference
Saccharum sp.	Ca	10% DMSO, 8% Glc	-1°C/min to -10°C 15 min, then to -40°C 2 hours, then LN	Yes	Yes	Jian et al. (1987)
	Ca, Su	10% PEG, 8% Glc, 10% DMSO	-1–3°C/min to -30°C, then LN	Yes	Yes	Ulrich et al. (1979)
	Su	Preculture 0.33 M Sor, then 0.64 M DMSO, 0.5 M Sor	-0.5°C/min to -40°C, then LN	Yes	Yes	Gnanapragasam and Vasil (1990)
	ST	Preculture 1 day, then 3% alginate 2 days, then desiccate	LN immersion	60%	NA	Paulet et al. (1993)
Salix sp.	ST	Winter twigs precooled to -5°C	-5°C/hour stepwise to -15 to -30°C, then LN, slow thaw	Yes	NA	Saka (1960, 1965)
Secale cereale	Pr	Preculture 1.75 M EG, then 7 M EG, 0.88 M Sor, 6% BSA	LN immersion	Yes	NR	Langis and Steponkus (1989, 1990, 1991)
Setaria italica	Ca, Su	10% DMSO, 0.5 M Sor	-1°C/min stepwise to -36°C, then LN	Yes	Yes	Lu and Sun (1992)
Silene sp.	ST	10% DMSO, 3% Glc	-0.5°C/min to -40°C, then LN	82–100%	82–88%	Fukai et al. (1991b)
Solanum etuberosum	ST	Preculture 2 days, then 10% DMSO	-0.3°C/min to -40°C, then LN	42–69%	Yes	Towill (1981)
Solanum goniocalyx	ST	5–10% DMSO	LN immersion	20–63%	Yes	Grout and Henshaw (1980)
Solanum microdontum	ST	Preculture 1 day, then 10% DMSO 1 hour	LN immersion	76%	48%	Ward et al. (1993)
Solanum phureja	ST	10% DMSO	-0.2–0.3°C/min to -35°C, then LN	96%	30%	Towill (1984)
Solanum pinnatisectum	ST	Preculture 1 day, then 10% DMSO 1 hour	LN immersion	67%	6%	Ward et al. (1993)
Solanum tuberosum	Me	1 week preculture, then 5% DMSO, 5% glycerol, 5% Suc	LN immersion	Yes	NA	Bajaj (1981a)
	ST	10% DMSO	LN immersion	Yes	NR	Harding et al. (1991)
Sorghum bicolor	Su	0.5 M glycerol, 0.5 M DMSO, 1 M Pro or Suc	-1°C/min to -35°C, hold 35 min, then LN	25–30%	NR	Withers and King (1980)
Spirea sp.	ST	Preculture 4% DMSO 2 days, then 8% DMSO, 8% Suc, or 12% DMSO	-0.25°C/min to -35°C, then LN	23%	NA	Towill (1991)
Theobroma cacao	ZE	Preculture 3% Suc, then 0.5 M Suc, 10% DMSO	-0.4°C/min to -35°C, then LN	Yes	2°	Pence (1991a)
Trifolium repens	Me	Preculture 5% DMSO, 5% Glc 4°C 2 days, then 10% DMSO, 10% Glc	-0.3°C/min to -40°C, then LN	>80%	Yes	Yamada et al. (1991)

Species	Tissue	Cryoprotectant/pretreatment	Freezing method	Survival	Regeneration	Reference
Triticum aestivum	PE	Preculture 4–6 weeks, then 5% DMSO, 5% Suc, 5% glycerol	LN immersion	19%	Yes	Bajaj (1983b)
	Pr	5% DMSO, 5% Glc	−1.3°C/min to −35°C, then LN	33%	Yes	Takeuchi *et al.* (1982)
	Su, Ca	5–15% DMSO, 0.5 M Suc	−0.5°C/min to −35°C, then LN	Yes	2°	Chen *et al.* (1985)
	ZE	10% PEG, 8% Glc, 10% DMSO stepwise	LN immersion	70%	NR	Zavala and Sussex (1986)
	ZE	Preculture 0.5 mg/liter ABA 10 days	−1°C/min to −35°C, then LN	Yes	Yes	Kendall *et al.* (1993)
Triticum aestivum × *Pisum sativum*	Pr	0.5 M Man, 5% DMSO	LN vapors	15%	NR	Bajaj (1983d)
Ulmus americana	Ca	10% PEG, 8% Glc, 10% DMSO	−1°C/min to −30°C, then LN	42%	Yes	Ulrich *et al.* (1984)
Vaccaria pyramidata	ST	10% DMSO, 3% Glc	−0.5°C/min to −40°C, then LN	100%	78%	Fukai *et al.* (1991b)
Vaccinium	Me	Cold harden −1°C nights, 3–7 wks	−1°C/min to −35°C, then LN	Yes	Yes	Reed (1989)
Veitchia merrillii	ZE	Desiccation to 10% H$_2$O	LN immersion	63–64%	Yes	Chin *et al.* (1988)
Vinca minor	Su	15% Suc, 5% glycerol	Slow to −80°C, then LN	Yes	NR	Caruso *et al.* (1988)
Vinca rosea	Su, Pr	Glc–Suc–Fru or 0.55 M Man	To −2°C, nucleated. −5°C/hour	Yes	NR	Bartolo *et al.* (1987)
Vitis sp.	Su	5% DMSO, 0.25 M maltose	−0.5°C/min to −40°C, then LN	Yes	NR	Dussert *et al.* (1991)
Zea mays	Su	0.5 M glycerol, 0.5 M Pro, 1 M DMSO 0°C 1 hour	−0.5°C/min to −40°C, then LN	Yes	Yes	DiMaio and Shillito (1989)
	ZE	On-ear desiccation or high osmoticum, preculture 15% Suc, then 5% DMSO, 5% glycerol	−1°C/min to −40°C, then LN	Yes	Yes	Delvallee *et al.* (1989)

[a] Abbreviations: An, anther; Ca, callus; EA, embryo axis; EC, embryonic cotyledon; LS, leafy section; Me, shoot meristem; Mi, microspore; Ov, ovule; PE, pollen embryo; Pr, protoplast; SE, somatic embryo; ST, shoot tip; Su, cell suspension; ZE, zygotic embryo.

[b] ABA, abscisic acid; BSA, bovine serum albumin; DMSO, dimethyl sulfoxide; EG, ethylene glycol; Glc, glucose; Man, mannitol; PEG, polyethylene glycol; Pro, proline; Sor, sorbitol; Suc, sucrose.

[c] LN, Liquid nitrogen.

[d] NA, Not applicable; NR, not reported; 2°, secondary (indirect) regeneration.

Freezing   Three basic methods have been employed for freezing plant materials: very rapid freezing, slow freezing, and droplet freezing (Kartha, 1982, 1987). Generally, rapid freezing is inferior to slow freezing, but it has been used successfully in the preservation of some cold-hardened shoots of woody species (Sakai, 1960, 1965).

With slow freezing, ice nucleation begins in the suspending medium and extracellular fraction. At this point, intracellular freezing is prevented by the cell membrane. As the extracellular fraction freezes, extracellular solutes become more concentrated, causing the cell to dehydrate, thus minimizing damaging intracellular ice formation (Bajaj, 1985; Kartha, 1987). The actual success of this freezing method, however, depends on the starting material, pretreatments and cryoprotectants, cooling rate, and terminal transfer temperature to liquid nitrogen (Kartha, 1987). Generally, a freezing rate of -0.25 to $-2°C$/min down to -30 to $-40°C$ has been effective for a large number of species. It is recommended, however, that the cooling rate and terminal transfer temperature optimum should be worked out for each species in question to balance the opposing factors of dehydration injury versus lethal intracellular ice formation (Henshaw, 1984).

Droplet freezing (placing the explants in a drop of liquid on a piece of aluminum foil, rather than in a vial or straw) has been effective for freezing some shoot explants, presumably by conferring a more homogeneous freezing of the tissues through a combination of smaller sample sizes and more efficient thermal conductivity between the sample and the liquid nitrogen (Kartha *et al.*, 1982).

Thawing   The vast majority of successful preservation protocols use a very rapid thawing method, in order to prevent lethal intracellular ice crystal growth (Kartha, 1981). A typical method is to plunge the frozen sample in a water bath held at 35–40°C unitl the ice is converted to liquid water (Bajaj, 1976a). Other methods have been used, such a higher thawing temperatures and microwave irradiation, to affect a more homogeneous warming of the sample (Reuff *et al.*, 1988).

Viability Testing   The most useful method for determining postpreservation viability is based on culture regrowth (Withers, 1986). Several rapid tests, however, have been employed to estimate viability, including triphenyltetrazolium chloride exclusion (Steponkus and Lanphear, 1967), infrared spectroscopy (Sowa and Towill, 1991b), fluorescein diacetate and phenosafranine (Widholm, 1972), crystal violet (Clulow *et al.*, 1991), trypan blue (Weber and Lark, 1979), and Evan's blue staining (Withers, 1984). To increase reliability, more than one test should be used (Kartha, 1987).

Recovery   Some studies have advocated the removal of cryoprotectants after thawing (Bajaj, 1976b). However, in some studies increased survival was obtained without washing to remove cryoprotectants (Cella *et al.*, 1982). Several studies recommend placing the thawed sample directly onto a semisolid medium to allow for a gradual dilution of cryoprotectants away from the sample (Withers, 1980; Seitz, 1987). The composition of the recovery medium is an important factor in promoting satisfactory recovery after freezing (Benson and Withers, 1988). Culturing thawed samples on a feeder layer to improve recovery has been reported (Hauptmann and Widholm, 1982). Organized cultures (shoot tips, meristems, and embryos) should be examined for the resumption of organized growth and regenerable lines should be examined for retention of regenerative capacity after cryopreservation (Kartha, 1981).

Evidence suggests that genetic stability is maintained in cryopreserved materials, and any genetic damage that may occur probably occurs during the actual freezing and thawing phases, rather than during storage (Benson and Withers, 1987; Benson and Harding, 1990). Exposure to background levels of radiation during storage may produce genetic change, because DNA repair mechanisms do not function at cryogenic temperatures; however, this is not believed to be of significant consequence for the practical storage times being envisioned (Towill, 1988; Towill and Roos, 1989).

Techniques useful in the characterization and evaluation of cultures for genetic stability include nuclear cytology (D'Amato, 1975), isozyme analysis (Simpson and Withers, 1986), fourth-derivative visible spectroscopy (Daley *et al.*, 1986), and restriction fragment-length polymorphism mapping (Helentjaris *et al.*, 1985; Harding 1991), as well as other molecular and biochemical methods.

Alternative Storage Methods

EMBRYOS

Somatic and zygotic embryos have been suggested as useful propagules for preservation (Towill and Roos, 1989). For example, preservation of some racalcitrant species has been made possible by the observation that excised embryos behave in an orthodox manner and can be cryopreserved (Grout, 1986; Williams, 1989; de Boucaud *et al.*, 1991).

Research has also been conducted to determine the feasibility of using desiccated somatic embryos (Gray, 1987) or encapsulated somatic embryos (Kitto and Janick, 1985; Dereuddre *et al.*, 1991a,b; Fabre and Dereuddre, 1990; Redenbaugh, 1990; Plessis *et al.*, 1991). Preservation of these "synthetic seeds" would be especially useful for preservation of clonal lines; however, more research is needed to elucidate how to increase viability after drying and to inhibit precocious germination.

NUCLEIC ACIDS

The total genetic information of a plant can be readily isolated and DNA segments can be stored in lyophilized form (Withers, 1987a). These genes can then be utilized in transformation experiments, either by *Agro-bacterium tumefaciens,* microprojectile bombardment, microinjection, electroporation, or other technologies (Peacock, 1989). Complete evaluation of the species, however, would not be possible (Towill and Roos, 1989), but this technique may be useful as a supplement to other, more complete storage methods.

POLLEN

Pollen storage can preserve genes, but may not preserve desirable gene combinations. Thus, it would be a useful method for storage of genes of interest to gene transfer technologies, similar to stored DNA. Pollen storage also would be a useful adjunct for germplasm preservation of lines developed for breeding programs; however, cytoplasmic genes may not be conserved (Bajaj, 1987; Towill and Roos, 1989).

Similar to seed, pollen can be classified as desiccation tolerant or desiccation sensitive (Towill and Roos, 1989), and species with disiccation-sensitive seed need not produce desiccation-sensitive pollen. Several studies have demonstrated survival of desiccation-tolerant pollen after short-term storage in liquid nitrogen (Ganeshan, 1985, 1986; Parfitt and Almehdi, 1984; Haunold and Stanwood, 1985; Towill, 1985, 1987; Hughes *et al.,* 1991). Storage of pollen in organic solvents has also been examined (Jain and Shivanna, 1988). Pollen viability can be routinely assessed on the basis of a fluorochromatic reaction (FCR) test (Shivanna *et al.,* 1991).

Information Sources

International Agricultural Research Centers

Many of the critical issues pertaining to long-term species preservation are applicable to *in vitro* preservation, and much information can be gained from their research programs and initiatives.

In 1974, the Consultative Group on International Agricultural Research (CGIAR) established the IBPGR to conserve, protect, and document germplasm resources of crop species through a worldwide network of gene banks (Williams, 1989; Sattaur, 1989; Shell, 1990; Gibbons, 1991). International Agricultural Research Centers with plant germplasm holdings include Centro Internacional de Agricultura Tropical (CIAT, Columbia), Centro Internacional de Mejoramiento de Maiz Trigo (CIMMYT, Mexico), Centro Internacional de la Papa (CIP, Peru), International Centre for Agricultural Research in Dry Areas (ICARDIA, Syria), International Crops Research Institute for the Semi-Arid Tropics (ICRISAT, India), International Institute of Tropical Agriculture (IITA, Nigeria), International Livestock Centre for Africa (ILCA, Ethiopia), International Rice Research Institute (IRRI, Philippines), and the West Africa Rice Development Association (WARDA, Cote d'Ivoire) (IBPGR, 1989).

A recent Conservation Database search revealed that 6165 unique species are conserved worldwide in the form of seeds, cultures, or field gene banks (L. A. Withers, personal communication). Since 1983, the IBPGR has sponsored a continuing information project at the University of Nottingham and has established an International Data Base on *In Vitro* Conservation to assist researchers in gathering information on institutes working on tissue culture techniques (Wheelans and Withers, 1984, 1988; Withers *et al.,* 1990). The IBPGR has a continuing mandate to fund germplasm preservation research projects (Coulman, 1988) and encourages the use of *in vitro* cultures for active collections. It has established *in vitro* working collections for cassava at CIAT, for potato at CIP, and for sweet potato at IITA (Staritsky *et al.,* 1986). Additionally, IBPGR has published recommendations for the operation of seed storage facilities (IBPGR, 1985) and *in vitro* collections, and has documented several important issues, such as genetic stability (Snowcroft, 1984), storage economy, and international exchange (IBPGR, 1986).

United States National Plant Germplasm System

The United States National Plant Germplasm System (NPGS) is a cooperating network of federal, state, and private-sector agencies, institutions, and research units involved in the collection, identification, preservation, evaluation, documentation, and distribution of agricultural and industrial crop species. Its three major operational functions are the working collections, the base collections, and germplasm services (Shands *et al.,* 1989). The bulk of the working collections is held in four Regional Plant Introduction Stations (Griffin, Georgia; Ames, Iowa; Geneva, New York; and Pullman, Washington) and eight National Clonal Germplasm Repositories (Brownwood, Texas; Corvallis, Oregon; Davis, California; Geneva, New York; Hilo, Hawaii; Miami, Florida/Mayaguez, Puerto Rico; Riverside, California; and Washington, D.C.). For specific genera assigned to each, the reader is directed to Shands *et al.,* (1989) and Hummer (1989). At present, the clonal germplasm repositories act as both working and base collections and some of these repositories are utilizing *in vitro* maintenance for selected species (Towill and Roos, 1989; Westwood, 1989). The National Seed Storage Laboratory in Fort Collins, Colorado, is the base collection for species preserved as seed, and in general it does not distribute plant materials.

The Germplasm Resources Information Network (GRIN) database (Beltsville, Maryland) compiles holdings information and can be accessed by anyone in the United States, Canada, Mexico, and the CGIAR Centers. Scientists may use this information to determine the existence of germplasm that may be useful for their research efforts (Mowder and Stoner, 1989a,b).

A secondary objective of the NPGS is to support and conduct reseach relating to improved methods for germplasm preservation. A recent search of the USDA Current Research Information System (CRIS) database revealed that research on germplasm preservation is being supported at United States Department of Agriculture stations and at Clemson University, Cornell University, Kentucky State University, Tuskegee University, The University of Missouri, and West Virginia University.

Culture Repositories

International Agricultural Research Centers

    Centro Internacional de Agricultura Tropical (CIAT)
    (International Center for Tropical Agriculture)
    Apartado Aereo 6713, Cali, Columbia
    Phone (57-23) 675050 or (57-23) 689343
    Fax (57-23) 647243
    Telex 396-05769 CIAT CO
    E-mail 157:CGI301
    Cable CINATROP
        (*Phaseolus* bean, cassava, rice, forage crops)

    Centro Internacional de Mejoramiento de Maiz y Trigo (CIMMYT)
    (International Maize and Wheat Improvement Center)
    P.O. Box 6-641, Mexico 06600, D.F. Mexico
    Phone (52-595) 42100 or (52-5) 761-3311
    Fax (52-595) 41069 or 43097
    Telex 1772023 CIMTME
    E-mail 157:CGI201
    Cable CENCIMMYT
        (wheat, maize, triticale)

    Centro Internacional de la Papa (CIP)
    (International Potato Center)
    Apartado 5969, Lima, Peru
    Phone (51-14) 366920
    Fax (51-14) 351570
    Telex (394) 25672 PE
    E-mail 157:CGI801 and 157:CGI043
    Cable CIPAPA
        (potato, sweet potato)

International Board for Plant Genetic Resources (IBPGR)
c/o Food and Agriculture Organization of the United Nations
Via delle Sette Chiese 142
00145 Rome, Italy
Phone (39-6) 574-4719
Fax (39-6) 5750309
Telex 4900005332 (IBR UI) via USA
E-mail 157:CGI101
Cable FOODAGRI

International Centre for Agricultural
  Research in the Dry Areas (ICARDA)
P.O. Box 5466, Aleppo, Syrian Arab Republic
Phone (963-21) 213433/213477/234890
Fax (963-21) 225105 or 213490
Telex (492) 331206/331208/331263 ICARDA SY
  (barley, lentil, faba bean, wheat chickpea)

International Council for Research in Agroforestry (ICRAF)
P.O. Box 30677, Nairobi, Kenya
Phone (254-2) 521450
Fax (254-2) 521001
Telex (987) 22048
E-mail 157:CGI236

International Crops Research Institute
  for the Semi-Arid Tropics (ICRISAT)
Pantancheru P.O., Andhra Pradesh, 502 324, India
Phone (91-842)224016
Fax (91-842) 241239
Telex 422203 or 4256366 ICRI IN
E-mail 157:CGI505
Cable CRISAT Hyderabad
  (sorghum, millet, chickpea, pigeonpea, groundnut)

International Institute of Tropical Agriculture (IITA)
PMB 5320, Ibadan, Nigeria
Phone (234-22) 400300-317
Fax (234-1) 611896 or (229) 301466 via IITA Benin
Telex TROPIB NG (905) 31417, 31159 or TDS IBA NG (905) 20311 (Box 015)
E-mail 10074:CGU018
Cable TROPFOUND, IKEJA
  (cassava, maize, plantain, cowpea, soybean, rice, yam)

International Livestock Centre for Africa (ILCA)
P.O. Box 5689, Addis Ababa, Ethiopia
Phone (251-1) 613215
Fax (251-1) 611892
Telex 980-21207 ILCA ET
E-mail 157:CGI070
Cable ILCAF
  (grasses, legumes)

International Network for the Improvement of Banana and Plantain (INIBAP)
Parc Scientifique Agropolis
Bât 7—Bd de la Lironde
34980 Montferrier-sur-Lez, France
Phone (33-67) 611302
Fax (33-67) 610334
Telex 490 376 INIBAP F

International Rice Research Institute (IRRI)
P.O. Box 933, Manila, Philippines
Phone (63-2) 818-1926 (Trunk Hunting Line)
Fax (63-2) 817-8470 or (63-2) 818-2087
Telex (ITT) 45365 RICE INST PM, (ITT) 40890 RICE PM (Los Banos), (RCA)
   22456 IRI PH, (EASTERN) 63786 RICE PN, CAPWIRE 14861 IRRI PS
   (Mail Box)
E-mail 157:CGI401
Cable RICEFOUND, MANILA

West Africa Rice Development Association (WARDA)
01 B.P. 2551, Bouaké 01, Côte d'Ivoire
Phone (225) 632395/634514/633242/632396
Fax (225) 634714
Telex 69138 ADRAO CI, BOUAKE
E-mail 157:CGI125
Cable ADRAO BOUAKE CI

United States National Plant Germplasm System

National Clonal Germplasm Repository—Brownwood
USDA—ARS
W.R. Poage Pecan Field Station
701 Woodson Road
Brownwood, TX 76801 USA
Phone (409) 272-1402
Fax (409) 272-1401
E-mail a031 cbrownwo
   (Carya, Castanea)

National Clonal Germplasm Repository—Corvallis
33447 Peoria Road
Corvallis, OR 97333 USA
Phone (503) 757 4448
Fax (503) 757-4548
   (Corylus, Fragaria, Humulus, Mentha, Pyrus, Ribes, Rubus, Vaccinium)

National Clonal Germplasm Repository—Davis
University of California
Davis, CA 95616-8607 USA
Phone (916) 752-6504

Fax (916) 752-2132
(Actinidia, Diospyros, Ficus, Juglans, Morus, Olea, Pistacia, Prunus, Punica, Vitis)

National Clonal Germplasm Repository—Geneva
New York State Agricultural Experiment Station
Cornell University
Geneva, NY 14456-0462 USA
Phone (315) 787-2244 or 787-2333
(Malus, Vitis)

National Clonal Germplasm Repository—Hilo
USDA—ARS
P.O. Box 4487
Hilo, HI 96720 USA
Phone (808) 959-5833
Fax (808) 959-3539
   (Ananas, Annona, Artocarpus, Averrhoa, Bactris, Canarium, Carica, Litchi, Macadamia, Malpighia, Nephelium, Passiflora, Psidium)

National Clonal Germplasm Repository—Miami
13601 Old Cutler Road
Miami, FL 33158 USA
Phone (305) 238-9321
   (Annona, Citrus, Coffea, Mangifera, Musa, Palmae, Persea, Saccharum, Theobroma, Tripsacum)

National Clonal Germplasm Repository for Citrus/Dates
1060 Pennsylvania Avenue
Riverside, CA 92507 USA
Phone (714) 787-4399
Fax (714) 787-4398
   (Citrus, Phoenix)

U.S. National Arboretum
3501 New York Avenue, N.E.
Washington, D.C. 20002 USA
Phone (202) 475-4836
Fax (202) 475-5252
   (woody ornamentals)

U.S. National Seed Storage Laboratory
Ft. Collins, CO 80523 USA
Phone (303) 484-0402
Fax (303) 221-1427

Database Management Unit
Germplasm Resources Information Network (GRIN)
USDA—ARS, Plant Sciences Institute
Bldg. 003, 4th floor, BARC-West
10300 Baltimore Avenue

Beltsville, MD 20705-2350 USA
Phone (301) 504-5666

Acknowledgments

The author thanks the U.S. National Seed Storage Laboratory and Clonal Germplasm Repositories, L.A. Withers of IBPGR, P.A. Sachariat and C.J. Britton of the OARDC Library, and W.K. Alley for their valuable assistance.

References

Abdelnour-Esquivel, A., Mora, A., and Villalobos, V. (1992a). *Cryo-Letters* **13**, 159–164.
Abdelnour-Esquivel, A., Villalobos, V., and Englemann, F. (1992b). *Cryo-Letters* **13**, 297–302.
Ashwood-Smith, M. J. (1985). *Cryobiology* **22**, 427–433.
Assy-Bah, B., and Engelmann, F. (1992). *Cryo-Letters* **13**, 117–126.
Augereau, J. M., Courtois, D., and Petiard, V. (1985). *In* "Aspects Industriels des Cultures Cellulaires d'Origine Animale et Végétale, 10th Coll. Soc. Fr. Microbiol., pp. 353–366. Lyon.
Augereau, J. M., Courtois, D., and Petiard, V. (1986). *Plant Cell Rep.* **5**, 372–376.
Bagniol, S., Engelmann, F., and Michaux-Ferrière, N. (1992). *Cryo-Letters* **13**, 405–412.
Bajaj, Y. P. S. (1976a). *Acta Hortic.* **63**, 75–84.
Bajaj, Y. P. S. (1976b). *Physiol. Plant.* **37**, 263–268.
Bajaj, Y. P. S. (1977). *Crop Improv.* **4**, 198–204.
Bajaj, Y. P. S. (1978). *Phytomorphology* **28**, 171–176.
Bajaj, Y. P. S. (1979). *Indian J. Exp. Biol.* **17**, 1405–1407.
Bajaj, Y. P. S. (1980). *Cereal Res. Commun.* **8**, 365–369.
Bajaj, Y. P. S. (1981a). *Euphytica* **30**, 141–145.
Bajaj, Y. P. S. (1981b). *Sci. Hortic. (Amsterdam)* **14**, 93–95.
Bajaj, Y. P. S. (1982). *Curr. Sci.* **51**, 139–140.
Bajaj, Y. P. S. (1983a). *Field Crops Res.* **7**, 161–167.
Bajaj, Y. P. S. (1983b). *In* "Proceedings of the Sixth International Wheat Genetics Symposium" (S. Sakamoto, ed.), pp. 565–574. Plant Germ-Plasm Inst., Fac. Agric., Kyoto.
Bajaj, Y. P. S. (1983c). *Curr. Sci.* **52**, 484–486.
Bajaj, Y. P. S. (1983d). *Indian J. Exp. Biol.* **21**, 120–122.
Bajaj, Y. P. S. (1984). *Curr. Sci.* **53**, 1215–1216.
Bajaj, Y. P. S. (1985). *In* "Cryopreservation of Plant Cells and Organs" (K. K. Kartha, ed.), pp. 227–242. CRC Press, Boca Raton, FL.
Bajaj, Y. P. S. (1987). *Int. Rev. Cytol.* **107**, 397–420.
Bajaj, Y. P. S. (1988). *Indian J. Exp. Biol.* **26**, 289–292.
Bakas, L. S., and Disalvo, E. A. (1991). *Cryobiology* **28**, 347–353.
Bartolo, M. E., Wallner, S. J., and Ketchum, R. E. (1987). *Cryobiology* **24**, 53–57.
Benson, E. E., and Harding, K. (1990). *In* "Plant Aging: Basic and Applied Approaches" (Rodriguez *et al.*, eds.), pp. 125–131. Plenum, New York.
Benson, E. E., and Noronha-Dutra, A. A. (1988). *Cryo-Letters* **9**, 120–131.
Benson, E. E., and Withers, L. A. (1987). *Cryo-Letters* **8**, 35–46.
Benson, E. E., and Withers, L. A. (1988). *In* "Plant Cell Biotechnology" (M. S. S. Pais, F. Mavituna, and J. M. Novais, eds.), pp. 431–443. Springer-Verlag, Berlin.

Bertrand-Desbrunais, A., Fabre, J., Engelmann, F., Dereuddre, J., and Charrier, A. (1988). *C. R. Seances Acad. Sci.* **307,** 795–801.

Bertrand-Desbrunais, A., Noirot, M., and Charrier, A. (1991). *Plant Cell, Tissue Organ Cult.* **27,** 333–339.

Bhandal, I. S., Hauptmann, R. M., and Widholm, J. M. (1985). *Plant Physiol.* **78,** 430–432.

Binder, W. D., and Zaerr, J. B. (1980a). *Cryobiology* **17,** 624.

Binder, W. D., and Zaerr, J. B. (1980b). *Cryobiology* **17,** 624–625.

Braun, A. (1988). *Plant Cell, Tissue Organ Cult.* **14,** 161–168.

Bridgen, M. P., and Staby, G. L. (1981). *Plant Sci. Lett.* **22,** 177–186.

Butenko, R. G., Popov, A. S., Volkova, L. A., Chernyak, N. D., and Nosov, A. M. (1984). *Plant Sci. Lett.* **33,** 285–292.

Caplin, S. M. (1959). *Am. J. Bot.* **46,** 324–329.

Caruso, M., Crespi-Perellino, L., Garofano, L., and Guicciardi, A. (1988). *In* "Plant Cell Biotechnology" (M. S. S. Pais, F. Mavituna, and J. M. Novais, eds.), pp. 271–274. Springer-Verlag, Berlin.

Cella, R., Colombo, R., Galli, M. G., Nielsen, E., Rollo, F., and Sala, F. (1982). *Physiol. Plant.* **55,** 279–284.

Charne, D. G., Pukacki, P., Kott, L. S., and Beversdorf, W. D. (1988). *Plant Cell Rep.* **7,** 407–409.

Chaudhury, R., Radhamani, J., and Chandel, K. P. S. (1991). *Cryo-Letters* **12,** 31–36.

Chen, T. H. H., and Li, P. H. (1989). *In* "Low Temperature Stress Physiology in Crops" (P. H. Li, ed.), pp. 139–152. CRC Press, Boca Raton, FL.

Chen, T. H. H., Kartha, K. K., and Gusta, L. V. (1985). *Plant Cell, Tissue Organ Cult.* **4,** 101–109.

Chin, H. F. (1988). "Recalcitrant Seeds: A Status Report." Int. Board Plant Genet. Resour., Rome.

Chin, H. F., and Roberts, E. H., eds. (1980). "Recalcitrant Crop Seeds." Malays. Trop. Press Sdn. Bhd., Kuala Lumpur.

Chin, H. F., Krishnapillay, B., and Alang, Z. C. (1988). *Cryo-Letters* **9,** 372–379.

Clulow, S. A., Wilkinson, M. J., Waugh, R., Baird, E., DeMaine, M. J., and Powell, W. (1991). *Theor. Appl. Genet.* **82,** 545–551.

Coulman, B. E. (1988). *Agrologist* **17,** 4–5.

Cromarty, A. S., Ellis, R. H., and Roberts, E. H. (1985). "The Design of Seed Storage Facilities for Genetic Conservation." Int. Board Plant Genet. Resour., Rome.

Daley, L. S., Thompson, M. M., Proebsting, W. M., Postman, J., and Jeong, B.-R. (1986). *Spectroscopy* **1,** 27–31.

D'Amato, F. (1975). *In* "Crop Genetic Resources for Today and Tomorrow" (O. H. Frankel and J. G. Hawkes, eds.), pp. 333–348. Cambridge Univ. Press, Cambridge, UK.

de Boucaud, M.-T., Brison, M., Ledoux, C., Germain, E., and Lutz, A. (1991). *Cryo-Letters* **12,** 163–166.

de Boucaud, M.-T., and Cambecedes J. (1988). *Cryo-Letters* **9,** 94–101.

Delvallee, I., Guillaud, J., Beckert, M., and Dumas, C. (1989). *Plant Sci.* **60,** 129–136.

Demeulemeester, M. A. C., Vandenbussche, B., and De Proft, M. P. (1993). *Cryo-Letters* **14,** 57–64.

Dereuddre, J., Galerne, M., and Gazeau, C. (1987). *C. R. Seances Acad. Sci., Ser. 3* **304,** (19), 485–487.

Dereuddre, J., Fabre, J., and Bassaglia, C. (1988). *Plant Cell Rep.* **7,** 170–173.

Dereuddre, J., Scottez, C., Arnaud, Y., and Duron, M. (1990a). *C. R. Seances Acad. Sci., Ser. 3* **310,** 265–272.

Dereuddre, J., Scottez, C., Arnaud, Y., and Duron, M. (1990b). *C. R. Seances Acad. Sci., Ser. 3* **310,** 317–323.

Dereuddre, J., Blandin, S., and Hassen, N. (1991a). *Cryo-Letters* **12,** 125–134.

Dereuddre, J., Hassen, N., Blandin, S., and Kaminski, M. (1991b). *Cryo-Letters* **12**, 135–148.

Diettrich, B., Wolf, T., Bormann, A., Popov, A. S., Butenko, R. G., and Luckner, M. (1987). *Planta Med.* **53**, 359–363.

Diettrich, B., Donath, P., Popov, A. S., Butenko, R. G., and Luckner, M. (1990). *Biochem. Physiol. Pflanz.* **186**, 63–67.

DiMaio, J. J., and Shillito, R. D. (1989). *J. Tissue Cult. Methods* **12**, 163–169.

Dussert, S., Mauro, M. C., Deloire, A., Hamon, S., and Engelmann, F. (1991). *Cryo-Letters* **12**, 287–298.

Ellis, R. H. (1984a). *Plant Genet. Resour. Newsl.* **58**, 16–33.

Ellis, R. H. (1984b). *In* "Seed Management Techniques for Genebanks: A Report of a Workshop Held 6-9th July 1982 at the Royal Botanic Gardens, Kew, U.K." (J. B. Dickie, S. Linington, and J. T. Williams, eds.), pp. 146–178. Int. Board Plant Genet. Resour., Rome.

Ellis, R. H., Hong, T. D., and Roberts, E. H. (1985). "Handbook of Seed Technology for Genebanks, Vol. 2. Int. Board Plant Genet. Resour., Rome.

Engelmann, F. (1990a). *Actual. Bot.* **137**, 93–98.

Engelmann, F. (1990b). *Int. J. Refrig.* **13**, 26–30.

Engelmann, F. (1992). *Cryo-Letters* **13**, 331–336.

Engelmann, F., and Dereuddre, J. (1988). *Cryo-Letters* **9**, 220–235.

Engelmann, F., Duval, Y., and Dereuddre, J. (1985). *C. R. Seances Acad. Sci., Ser. 3* **301**, 111–116.

Fabre, J., and Dereuddre, J. (1990). *Cryo-Letters* **11**, 413–426.

Fahy, G. M., Levy, D. I., and Ali, S. E. (1987). *Cryobiology* **24**, 196–213.

Farrant, J. M., Pammenter, N. W., and Berjak, P. (1988). *Seed Sci. Technol.* **16**, 155–166.

Finkle, B. J., and Ulrich, J. M. (1982). *Cryobiology* **19**, 329–335.

Finkle, B. J., Ulrich, J. M., Rains, D. W., and Stavarek, S. J. (1985a). *Plant Sci.* **42**, 133–140.

Finkle, B. J., Zavala, M. E., and Ulrich, J. M. (1985b). *In* "Cryopreservation of Plant Cells and Organs" (K. K. Kartha, ed.), pp. 75–113. CRC Press, Boca Raton, FL.

Ford, K. G. (1990). *Plant Cell Rep.* **8**, 534–537.

Ford-Lloyd, B. V. (1990). *Adv. Hortic. Sci.* **4**, 31–38.

Fowler, M. W., and Scragg, A. H. (1988). *In* "Plant Cell Biotechnology" (M. S. S. Pais, F. Mavituna, and J. M. Novais, eds.), pp. 165–177. Springer-Verlag, Berlin.

Friesen, L. J., Kartha, K. K., Leung, N. L., Englund, P., Giles, K. L., Park, J., and Songstad, D. D. (1991). *Planta Med.* **57**, 53–55.

Fukai, S. (1990). *Sci. Hort.* (*Amsterdam*) **45**, 167–174.

Fukai, S., Goi, M., and Tanaka, M. (1991a). *Euphytica* **54**, 201–204.

Fukai, S., Goi, M., and Tanaka, M. (1991b). *Euphytica* **56**, 149–153.

Ganeshan, S. (1985). *Vitis* **24**, 169–173.

Ganeshan, S. (1986). *Trop. Agric.* **63**, 46–48.

Gazeau, C. M., Blachon, C., and Dereuddre, J. (1992). *Cryo-Letters* **13**, 149–158.

Gibbons, A. (1991). *Science* **254**, 804.

Gnanapragasam, S., and Vasil, I. K. (1990). *Plant Cell Rep.* **9**, 419–423.

Gnanapragasam, S., and Vasil, I. K. (1992). *Plant Sci.* **83**, 205–215.

Gonzalez-Benito, M. E., and Perez-Ruiz, C. (1992). *Cryobiology* **29**, 685–690.

Gray, D. J. (1987). *HortScience* **22**, 810–814.

Grout, B. W. W. (1986). *In* "Plant Tissue Culture and its Agricultural Applications" (L. A. Withers and P. G. Alderson, eds.), pp. 303–309. Butterworth, London.

Grout, B. W. W., and Henshaw, G. G. (1980). *Ann. Bot.* (*London*) [N. S.] **46**, 243–248.

Grout, B. W. W., and Morris, G. J. (1987). *In* "The Effects of Low Temperatures on Biological Systems" (B. W. W. Grout and G. J. Morris, eds.), pp. 147–173. Edward Arnold, London.

Grout, B. W. W., Westcott, R. J., and Henshaw, G. G. (1978). *Cryobiology* **15**, 478–483.

Grout, B. W. W., Shelton, K., and Pritchard, H. W. (1983). *Ann. Bot. (London)* [N. S.] **52**, 381–384.

Gunning, J., and Lagerstedt, H. B. (1986). *Comb. Proceedings Int. Plant Prop. Soc.* **35**, 199–205.

Hahne, G., and Lorz, H. (1987). *Plant Breed.* **99**, 330–332.

Hanson, J. (1985). "Procedures for Handling Seeds in Genebanks." Int. Board Plant Genet. Resour., Rome.

Harada, T., Inaba, A., Yakuwa, T., and Tamura, T. (1985). *HortScience* **20**, 678–680.

Harding, K. (1991). *Euphytica* **55**, 141–146.

Harding, K., Benson, E. E., and Smith, H. (1991). *Cryo-Letters* **12**, 17–22.

Harrington, J. F. (1963). *Proc. Int. Seed Test. Assoc.* **28**, 989–994.

Haskins, R. H., and Kartha, K. K. (1980). *Can. J. Bot.* **58**, 833–840.

Haunold, A., and Stanwood, P. C. (1985). *Crop. Sci.* **25**, 194–196.

Hauptmann, R. M., and Widholm, J. M. (1982). *Plant Physiol.* **70**, 30–34.

Helentjaris, T., King, G., Slocum, M., Siedenstrang, C., and Wegman, S. (1985). *Plant Mol. Biol.* **5**, 109–118.

Henshaw, G. G. (1984). In "Crop Breeding: A Contemporary Basis" (P. B. Vose and S. G. Blixt, eds.), pp. 400–413. Pergamon, Oxford.

Hiraoka, N., and Kodama, T. (1984). *Plant Cell, Tissue Organ Cult.* **3**, 349–357.

Hughes, H. G., Lee, C. W., and Towill, L. E. (1991). *HortScience* **26**, 1411–1412.

Hummer, K. (1989). *HortScience* **24**, 190.

International Board for Plant Genetic Resources (IBPGR) (1985). "Cost-Effective Long-Term Seed Stores: A Report of the Meeting of a Sub-committee of the IBPGR Advisory Committee of Seed Storage, Held at Reading, UK, 1-3 April 1985." IBPGR, Rome.

International Board for Plant Genetic Resources (IBPGR) (1986). "Design, Planning and Operation of *in vitro* Genebanks: Report of a Subcommittee of the IBPGR Advisory Committee on *in vitro* Storage." IBPGR, Rome.

International Board for Plant Genetic Resources (IBPGR) (1989). "Partners in Conservation." IBPGR, Rome.

Jain, A., and Shivanna, K. R. (1988). *Ann. Bot. (London)* **61**, 325–330.

James, E. (1983). In "Plant Biotechnology" (S. H. Mantell and H. Smith, eds.), pp. 163–186. Cambridge Univ. Press, Cambridge, UK.

Jarret, R. L. and Florkowski, W. J. (1990). *HortScience* **25**, 141–146.

Jekkel, Z. S., Heszky, L. E., and Ali, A. H. (1989). *Acta Biol. Hung.* **40**, 127–136.

Jian, L. C., Sun, D. L., and Sun, L. H. (1987). *Plant Biol.* **5**, 323–327.

Jörgensen, J. (1991). In "Woody Plant Biotechnology" (M. R. Ahuja, ed.), pp. 355–356. Plenum, New York.

Justice, O. L., and Bass, L. N. (1978). "Principles and Practices of Seed Storage." U.S. Gov. Printing Office, Washington, DC.

Kartha, K. K. (1980). *J. Am. Soc. Hortic. Sci.* **105**, 481–484.

Kartha, K. K. (1981). In "Plant Tissue Culture Methods and Applications in Agriculture" (T. A. Thorpe, ed.), pp. 181–211. Academic Press, New York.

Kartha, K. K. (1982). In "Plant Cell & Tissue Culture" (A. Fujiwara, ed.), pp. 795–796. Jpn. Assoc. Plant Tissue Cult., Tokyo.

Kartha, K. K., ed. (1985). "Cryopreservation of Plant Cells and Organs." CRC Press, Boca Raton, FL.

Kartha, K. K. (1987). In "Cell Culture and Somatic Cell Genetics" (I. K. Vasil, ed.), Vol. 4, pp. 217–227. Academic Press, Orlando, FL.

Kartha, K. K., Leung, N. L., and Mroginski, L. A. (1982). *Z. Pflanzenphysiol.* **107**, 133–140.

Kartha, K. K., Fowke, L. C., Leung, N. L., Caswell, K. L., and Hakman, I. (1988). *J. Plant Physiol.* **132,** 529–539.

Katano, M., Ishihara, A., and Sakai, A. (1983). *HortScience* **18,** 707–708.

Kendall, E. J., Kartha, K. K., Qureshi, J. A., and Chermak, P. (1993). *Plant Cell Rep.* **12,** 89–94.

King, M. W., and Roberts, E. H. (1980a). *In* "Recalcitrant Crop Seeds" (H. F. Chin and E. H. Roberts, eds.), pp. 53–89. Malays. Trop. Press Sdn. Bhd., Kuala Lumpur.

King, M. W., and Roberts, E. H. (1980b). *In* "Recalcitrant Crop Seeds" (H. F. Chin, and E. H. Roberts, eds.), pp. 90–110. Malays. Trop. Press Sdn. Bhd., Kuala Lumpur.

Kitto, S. L., and Janick, J. (1985). *J. Am. Soc. Hortic. Sci.* **110,** 283–286.

Klimaszewska, K., Ward, C., and Cheliak, W. M. (1992). *J. Exp. Bot.* **43,** 73–79.

Kuoksa, T., and Hohtola, A. (1991). *Plant Cell, Tissue Organ Cult.* **27,** 89–93.

Kuriyama, A., Watanabe, K., Ueno, S., and Mitsuda, H. (1990). *Cryo-Letters* **11,** 171–178.

Kwiatkowski, S., Martin, M. W., Brown, C. R., and Sluis, C. J. (1988). *Am. Potato J.* **65,** 369–375.

Lainé, E., Bade, P., and David, A. (1992). *Plant Cell Rep.* **11,** 295–298.

Langis, R., and Steponkus, P. L. (1989). *Cryobiology* **26,** 575.

Langis, R., and Steponkus, P. L. (1990). *Plant Physiol.* **92,** 666–671.

Langis, R., and Steponkus, P. L. (1991). *Cryo-Letters* **12,** 107–112.

Langis, R., Schnabel, R., Earle, E. D., and Steponkus, P. L. (1989). *Cryo-Letters* **10,** 421–428.

Lecouteux, C., Florin, B., Tessereau, H., Bollon, H., and Petiard, V. (1991). *Cryo-Letters* **12,** 319–328.

Li, P. H., ed. (1989). "Low Temperature Stress Physiology in Crops." CRC Press, Boca Raton, FL.

Lizarraga, R., Huaman, Z., and Dodds, J. H. (1989). *Am. Potato J.* **66,** 253–269.

Lu, T. G., and Sun, C. S. (1992) *J. Plant Physiol.* **139,** 295–298.

Maddox, A. D., Gonsalves, F., and Shields, R. (1982–1983). *Plant Sci. Lett.* **28,** 157–162.

Mannonen, L., Toivonen, L., and Kauppinen, V. (1990). *Plant Cell Rep.* **9,** 173–177.

Marin, M. L., Mafla, G., Roca, W. M., and Withers, L. A. (1990). *Cryo-Letters* **11,** 257–264.

Mathias, S. F., Franks, F., and Hatley, R. H. M. (1985). *Cryobiology* **22,** 537–546.

Matthes, G., and Heckensellner, K. D. (1981). *Cryo-Letters* **2,** 389–392.

Mazur, R. A., and Hartmann, J. X. (1979). *In* "Plant Cell and Tissue Culture: Principles and Applications" (W. R. Sharp, P. O. Larson, E. F. Paddock, and V. Raghavan, eds.), p. 876. Ohio State Univ. Press, Columbus.

Meijer, E. G. M., Van Iren, F., Schrijnemakers, E., Hensgens, L. A. M., and van Zijderveld, M. (1991). *Plant Cell Rep.* **10,** 171–174.

Mix, G. (1982). *Plant Genet. Resour. Newsl.* **51,** 6–8.

Moriguchi, T., and Yamaki, S. (1989). *HortScience* **24,** 372–373.

Moriguchi, T., Kozaki, I., Matsuta, N., and Yamaki, S. (1988). *Plant Cell, Tissue Organ Cult.* **15,** 67–71.

Mowder, J. D., and Stoner, A. K. (1989a). *Plant Breed. Rev.* **7,** 57–65.

Mowder, J. D., and Stoner, A. K. (1989b). *In* "Biotic Diversity and Germplasm Preservation: Global Imperatives" (L. Knutson and A. K. Stoner, eds.), pp. 419–426. Kluwer Academic Publishers, Dordrecht, The Netherlands.

Mullin, R. H., and Schlegel, D. E. (1976). *HortScience* **11,** 100–101.

Mycock, D. J., Watt, M. P., and Berjak, P. (1991). *J. Plant Physiol.* **138,** 728–733.

Nag, K. K., and Street, H. E. (1975a). *Physiol. Plant.* **34,** 254–260.

Nag, K. K., and Street, H. E. (1975b). *Physiol. Plant.* **34,** 261–265.

Ng, S. Y. C., and Ng, N. Q. (1991). *In* "*In Vitro* Methods for Conservation of Plant Genetic Resources" (J. H. Dodds, ed.), pp. 11–39.

Niino, T., and Oka, S. (1990). *J. Seric. Sci. Jpn.* **59,** 111–117.

Nishizawa, S., Sakai, A., Amano, Y., and Matsuzawa, T. (1992). *Cryo-Letters* **13,** 379–388.

Nitzsche, W. (1978). *Z. Pflanzenphysiol.* **87**, 469–472.

Normah, M. N., and Vengadasalam, M. (1992). *Cryo-Letters* **13**, 199–208.

Normah, M. N., Chin, H. F., and Hor, Y. L. (1986). *Pertanika* **9**, 299–304.

Oka, S., Yakuwa, H., Sato, K., and Niino, T. (1991). *HortScience* **26**, 65–66.

Pammenter, N. W., Vertucci, C. W., and Berjak, P. (1991). *Plant Physiol.* **96**, 1093–1098.

Panis, B. J., Withers, L. A., and de Langhe, E. A. L. (1990). *Cryo-Letters* **11**, 337–350.

Parfitt, D. E., and Almehdi, A. A. (1984). *Fruit Var. J.* **38**, 14–16.

Paulet, F., Engelmann, F., and Glaszmann, J.-C. (1993). *Plant Cell Rep.* **12**, 525–529.

Peacock, W. J. (1989). *In* "The Use of Plant Genetic Resources" (A. D. H. Brown, ed.), pp. 363–376. Cambridge Univ. Press, Cambridge, UK.

Pence, V. C. (1988). *In Vitro Cell. Dev. Biol.* **24**, 33A.

Pence, V. C. (1990). *Cryobiology* **27**, 212–218.

Pence, V. C. (1991a). *Plant Cell Rep.* **10**, 144–147.

Pence, V. C. (1991b). *Seed Sci. Technol.* **19**, 235–251.

Pence, V. C. (1992). *Cryobiology* **29**, 391–399.

Plessis, P., Leddet, C., and Dereuddre, J. (1991). *C. R. Seances Acad. Sci., Ser. 3* **313**, 373–380.

Pritchard, H. W., and Prendergast, F. G. (1986). *J. Exp. Bot.* **37**, 1388–1397.

Pritchard, H. W., Grout, B. W. W., and Short, K. C. (1986). *Ann. Bot. (London)* [N. S.] **57**, 379–387.

Radhamani, J., and Chandel, K. P. S. (1992). *Plant Cell Rep.* **11**, 372–374.

Redenbaugh, K. (1990). *HortScience* **25**, 251–255.

Reed, B. M. (1988). *Cryo-Letters* **9**, 166–171.

Reed, B. M. (1989). *Cryo-Letters* **10**, 315–322.

Reed, B. M. (1990a). *Fruit Var. J.* **44**, 141–148.

Reed, B. M. (1990b). *HortScience* **25**, 111–113.

Reed, B. M. (1991). *Plant Cell Rep.* **10**, 431–434.

Reed, B. M. (1992). *Fruit Var. J.* **46**, 98–102.

Reisch, B. I. (1988). *In* "Plant Cell Biotechnology" (M. S. S. Pais, F. Mavituna, and J. M. Novais, eds.), pp. 87–95. Springer-Verlag, Berlin.

Reuff, I., Seitz, U., Ulbbrich, B., and Reinhard, E. (1988). *J. Plant Physiol.* **133**, 414–418.

Roberts, E. H. (1975). *In* "Crop Genetic Resources for Today and Tomorrow" (O. H. Frankel and J. G. Hawkes, eds.), pp. 269–295. Cambridge Univ. Press, Cambridge, UK.

Roberts, E. H., and King, M. W. (1980). *In* "Recalcitrant Crop Seeds" (H. F. Chin and E. H. Roberts, eds.), pp. 1–5. Malays. Trop. Press Sdn. Bhd., Kuala Lumpur.

Roberts, E. H., King, M. W., and Ellis, R. H. (1984). *In* "Crop Genetic Resources: Conservation and Evaluation" (J. H. W. Holden and J. T. Williams, eds.), pp. 38–52. Allen & Unwin, London.

Roos, E. E. (1989). *Plant Breed. Rev.* **7**, 129–158.

Roos, E. E., and Wiesner, L. E. (1991). *HortTechnology* **1**, 65–69.

Sakai, A. (1960). *Nature (London)* **185**, 393–394.

Sakai, A. (1965). *Plant Physiol.* **40**, 882–887.

Sakai, A., and Nishiyamam, Y. (1978). *HortScience* **13**, 225–227.

Sakai, A., and Noshiro, M. (1975). *In* "Crop Genetic Resources for Today and Tomorrow" (O. H. Frankel and J. G. Hawkes, eds.), pp. 317–326. Cambridge Univ. Press, Cambridge, UK.

Sakai, A., and Otsuka, K. (1972). *Plant Cell Physiol.* **13**, 1129–1133.

Sakai, A., and Sugawara, Y. (1973). *Plant Cell Physiol.* **14**, 1201–1204.

Sakai, A., Kobayashi, S., and Oiyama, I. (1990). *Plant Cell Rep.* **9**, 30–33.

Sattaur, O. (1989). *New Sci.*, 29 July, pp. 37–41.

Schaper, D., and Zimmer, K. (1989). *Gartenbauwissenschaft* **54**, 85–89.

Seitz, U. (1987). *Planta Med.* **53**, 311–314.
Seitz, U., Alfermann, A. W., and Reinhard, E. (1983). *Plant Cell Rep.* **2**, 273–276.
Senft, D. (1989). *Agric. Res.* **37**, 21.
Shands, H. L., Fitzgerald, P. J., and Eberhart, S. A. (1989). *In* "Biotic Diversity and Germplasm Preservation: Global Imperatives" (L. Knutson and A. K. Stoner, eds.), pp. 95–115. Kluwer Academic Publishers, Dordrecht, The Netherlands.
Shell, E. R. (1990). *Smithsonian,* January, pp. 94–105.
Shillito, R. D., Carswell, G. K., Johnson, C. M., DiMaio, J. J., and Harms, C. T. (1989). *Bio/Technology* **7**, 581–587.
Shimonishi, K., Ishikawa, M., Suzuki, S., and Oosawa, K. (1991). *Jpn. J. Breed.* **41**, 347–351.
Shivanna, K. R., Linskens, H. F., and Cresti, M. (1991). *Theor. Appl. Genet.* **81**, 38–42.
Siminovitch, D. (1979). *Plant Physiol.* **63**, 722–725.
Simpson, M. J. A., and Withers, L. A. (1986). "Characterization of Plant Genetic Resources Using Isozyme Electrophoresis: A Guide to the Literature. Int. Board Plant Genet. Resour., Rome.
Snowcroft, W. R. (1984). "Genetic Variability in Tissue Culture: Impact on Germplasm Conservation and Utilization." Int. Board Plant Genet. Resour., Rome.
Sowa, S., and Towill, L. E. (1991a). *Plant Cell, Tissue/Organ Cult.* **27**, 197–201.
Sowa, S., and Towill, L. E. (1991b). *Plant Physiol.* **95**, 610–615.
Sowa, S., Roos, E. E., and Zee, F. (1991). *HortScience* **26**, 597–599.
Stanwood, P. C. (1980). *J. Seed Technol.* **5**, 26–31.
Stanwood, P. C. (1985). *In* "Cryopreservation of Plant Cells and Organs" (K. K. Kartha, ed.), pp. 199–226. CRC Press, Boca Raton, FL.
Stanwood, P. C., and Bass, L. N. (1978). *In* "Plant Cold Hardiness and Freezing Stress" (P. H. Li and A. Sakai, eds.), pp. 361–371. Academic Press, New York.
Stanwood, P. C., and Bass, L. N. (1981). *Seed Sci. Technol.* **9**, 423–437.
Staritsky, G., Dekkers, A. J., Louwaars, N. P., and Zandvoort, E. A. (1986). *In* "Plant Tissue Culture and its Agricultural Applications" (L. A. Withers and P. G. Alderson, eds.), pp. 277–283. Butterworth, London.
Steponkus, P. L., and Lanphear, F. O. (1967). *Plant Physiol.* **42**, 1423–1426.
Strauss, A., Fankhauser, H., and King, P. J. (1985). *Planta* **163**, 554–562.
Styles, E. D., Burgess, J. M., Mason, C., and Huber, B. M. (1982). *Cryobiology* **19**, 195–199.
Sugawara, Y., and Takeuchi, M. (1988). *Cryobiology* **25**, 517.
Syroedov, V. I., Nadykta, V. D., Nazarova, N. I., Ragulin, M. S., and Gorshkov, G. S. (1987). *Sov. Agric. Sci.* **9**, 27–30.
Takeuchi, M., Matsushima, H., and Sugawara, Y. (1982). *In* "Plant Tissue & Cell Culture" (A. Fujiwara, ed.), pp. 797–798. Jpn. Assoc. Plant Tissue Cult., Tokyo.
Taniguchi, K., Tanaka, R., Ashitani, N., and Miyagawa, H. (1988). *Jpn. J. Genet.* **63**, 267–272.
Toivonen, P. M. A., and Kartha, K. K. (1989). *J. Plant Physiol.* **134**, 766–768.
Touchell, D. H., Dixon, K. W., and Tan, B. (1992). *Aust. J. Bot.* **40**, 305–310.
Towill, L. E. (1981). *Plant Sci. Lett.* **20**, 315–324.
Towill, L. E. (1984). *Cryo-Letters* **5**, 319–326.
Towill, L. E. (1985). *In* "Cryopreservation of Plant Cells and Organs" (K. K. Kartha, ed.), pp. 171–198. CRC Press, Boca Raton, FL.
Towill, L. E. (1987). "Biotechnology in Agriculture and Forestry" (Y. P. S. Bajaj, ed.), Vol. 3, pp. 427–440. Springer-Verlag, Berlin.
Towill, L. E. (1988). *HortScience* **23**, 91–95.
Towill, L. E. (1989). *Plant Breed. Rev.* **7**, 159–182.
Towill, L. E. (1990). *Plant Cell Rep.* **9**, 178–180.

Towill, L. E. (1991). In *"In Vitro* Methods for Conservation of Plant Genetic Resources" (J. H. Dodds, ed.), pp. 41–70. Chapman & Hall, London.

Towill, L. E., and Jarret, R. L. (1992). *Plant Cell Rep.* **11**, 175–178.

Towill, L. E., and Roos, E. E. (1989). In "Biotic Diversity and Germplasm Preservation" (L. Knutson and A. K. Stoner, eds.), pp. 379–403. Kluwer Academic Publishers, Dordrecht, The Netherlands.

Tyler, N. J., and Stushnoff, C. (1988a). *Can. J. Plant Sci.* **68**, 1163–1167.

Tyler, N. J., and Stushnoff, C. (1988b). *Can. J. Plant Sci.* **68**, 1169–1176.

Ulrich, J. M., Finkle, B. J., Moore, P. H., and Ginoza, H. (1979). *Cryobiology* **16**, 550–556.

Ulrich, J. M., Finkle, B. J., and Tisserat, B. H. (1982). *Plant Physiol.* **69**, 624–627.

Ulrich, J. M., Mickler, R. A., Finkle, B. J., and Karnosky, D. F. (1984). *Can. J. For. Res.* **14**, 750–753.

Uragami, A., Sakai, A., Nagai, M., and Takahashi, T. (1989). *Plant Cell Rep.* **8**, 418–421.

Uragami, A., Sakai, A., and Nagai, M. (1990). *Plant Cell Rep.* **9**, 328–331.

Vertucci, C. W. (1989). *Plant Physiol.* **90**, 1478–1485.

Vertucci, C. W. (1992). *Plant Physiol.* **99**, 310–316.

Wanas, W. H., and Callow, J. A. (1986). In "Plant Tissue Culture and its Agricultural Applications" (L. A. Withers and P. G. Alderson, eds.), pp. 285–290. Butterworth, London.

Ward, A. C. W., Benson, E. E., Blackhall, N. W., Cooper-Bland, S., Powell, W., Power, J. B., and Davey, M. R. (1993). *Cryo-Letters* **14**, 145–152.

Weber, G., and Lark, K. G. (1979). *Theor. Appl. Genet.* **55**, 81–86.

Weber, G., Roth, E. J., and Schweiger, H.-G. (1983). *Z. Pflanzenphysiol.* **109**, 29–39.

Westwood, M. N. (1989). *Plant Breed. Rev.* **7**, 111–128.

Wheelans, S. K., and Withers, L. A. (1984). *Plant Genet. Res. Newsl.* **60**, 33–38.

Wheelans, S. K., and Withers, L. A. (1988). In "Plant Cell Biotechnology" (M. S. S. Pais, F. Mavituma, and J. M., Novais, eds.), pp. 497–500. Springer-Verlag, Berlin.

Widholm, J. M. (1972). *Stain Technol.* **47**, 189–194.

Wilkes, G. (1984). In "Plant Genetic Resources: A Conservation Imperative" (C. W. Yeatman, D. Kafton, and G. Wilkes, eds.), pp. 131–164. Westview Press, Boulder, CO.

Williams, J. T. (1989). In "Biotic Diversity and Germplasm Preservation: Global Imperatives" (L. Knutson and A. K. Stoner, eds.), pp. 81–96. Kluwer Academic Publishers, Dordrecht, The Netherlands.

Withers, L. A. (1980). *Cryo-Letters* **1**, 239–250.

Withers, L. A. (1984). In "Cell Culture and Genetics of Plants" (I. K. Vasil, ed.), Vol. 1, pp. 608–620. Academic Press, Orlando, FL.

Withers, L. A. (1985a). In "Cell Culture and Somatic Cell Genetics of Plants" (J. K. Vasil, ed.), Vol. 2, pp. 253–316. Academic Press, Orlando, FL.

Withers, L. A. (1985b). In "Plant Cell Culture: A Practical Approach" (R. A. Dixon, ed.), pp. 169–191. IRL Press, Oxford.

Withers, L. A. (1986). In "Plant Tissue Culture and its Agricultural Applications" (L. A. Withers and P. G. Alderson, eds.), pp. 261–276. Butterworth, London.

Withers, L. A. (1987a). *Oxford Surv. Plant Mol. Biol.* **4**, 221–272.

Withers, L. A. (1987b). In "The Effects of Low Temperatures on Biological Systems" (B. W. W. Grout and G. J. Morris, eds.), pp. 389–409. Edward Arnold, London.

Withers, L. A. (1990). *Methods Mol. Biol.* **6**, 39–48.

Withers, L. A. (1991a). *Biol. J. Linn. Soc.* **43**, 31–42.

Withers, L. A. (1991b). In "Maintenance of Microorganisms and Cultured Cells: A Manual of Laboratory Methods" (B. E. Kirsop and A. Doyle, eds.), 2nd ed., pp. 243–267. Academic Press, San Diego, CA.

Withers, L. A., and King, P. J. (1980). *Cryo-Letters* **1**, 213–220.

Withers, L. A., and Street, H. E. (1977). *Physiol. Plant.* **39,** 171–178.
Withers, L. A., Benson, E. E., and Martin, M. (1988). *Cryo-Letters* **9,** 114–119.
Withers, L. A., Wheelans, S. K., and Williams, J. T. (1990). *Euphytica* **45,** 9–22.
Woodstock, L. W., Maxon, S., and Bass, L. (1983). *J. Am. Soc. Hortic. Sci.* **108,** 692–696.
Yakuwa, H., and Oka, S. (1988). *Ann. Bot.* (*London*) [N. S.] **62,** 79–82.
Yamada, T., Sakai, A., Matsumura, T., and Higuchi, S. (1991). *Plant Sci.* **73,** 111–116.
Zavala, M. E., and Sussex, I. M. (1986). *J. Plant Physiol.* **122,** 193–197.
Zee, F. T., and Munekata, M. (1992). *HortScience* **27,** 57–58.
Zenk, M. H. (1978). *In* "Frontiers of Plant Tissue Culture 1978" (T. A. Thorpe, ed.), pp. 1–14. University of Calgary, Calgary, Canada.

CHAPTER 11

Plant Viruses and Viroids

Rose W. Hammond

J. Hammond

Background

Introduction

Diseases of plants known to be caused by viruses have been recognized and studied since the late nineteenth century. In contrast, plant diseases attributed to viroids have only recently been discovered. In a mature virus particle, the genetic material is usually encased in a protective protein coat or lipoprotein; viroids, on the other hand, are naked nucleic acid molecules. Although viruses and viroids are distinct classes of plant pathogens, methods used for their preservation are quite similar. Reliable stock cultures are required for a variety of reasons, including the maintenance of reference strains and pure cultures for research. For these reasons and others the American Type Culture Collection (ATCC) maintains a diverse collection of plant viruses and viroids (McDaniel, 1989).

In this chapter, we will first give a brief description of the features of plant viruses and viroids, including modes of transmission and methods used for characterization, followed by classical methods used in long-term preservation. Finally, we will discuss the use of cloned, infectious cDNA copies of viruses and viroids as an alternate means of preservation. It is the object of this chapter to provide a reference source for a variety of techniques that can be used in a laboratory for the maintenance and preservation of these particular pathogens. We stress that it is not all inclusive and refer the reader to the cited literature for more complete information.

Classification and Diversity

VIRUSES

There are currently 32 plant virus groups and three families of plant viruses as approved by the International Committee for the Nomenclature of Viruses (Martelli, 1992) (Table 1), with 300 members and 270 potential members (including strains and isolates). The majority of plant viruses (over 70%) consist of genomes of positive-sense, single-stranded RNA molecules, which can act directly as a messenger RNA when infecting the cell (Matthews, 1991). Plant viral genomes may also be composed of double-stranded, circular DNA; single-stranded, circular DNA; double-stranded RNA; single-stranded negative-sense RNA; and single-stranded ambisense RNA. In addition, particle morphology ranges from long, flexous rods to small isometrics, to more complex particles. In cases when the genetic material consists of more than one molecule, these molecules may each be enclosed, or encapsidated, within separate particles, or may all be present in the same particle. The genomes of plant viruses vary widely in size and the number of proteins encoded by their genomes. Purified virus preparations from infected plants may also contain a variety of RNAs other than the subgenomic RNAs. These include satellite RNAs and defective interfering RNAs (Matthews, 1991). The replication of these RNAs is dependent on the helper virus and can also affect disease symptoms caused by the virus in some host plants—either ameliorating or intensifying symptoms. Plant viruses, therefore, represent a very diverse group of plant pathogens. This diversity is many times reflected in the most effective methods used for their long-term storage.

VIROIDS

Many viruslike diseases of plants have been recently attributed to viroids. Viroids differ from conventional viruses in being free RNAs that lack apparent mRNA activity (Diener, 1987). They do not require a helper virus for replication and movement. The single-stranded, covalently closed circular RNA genomes of known viroids range in size from 246 to 375 nucleotides. Currently, over 16 plant diseases are known to be caused by viroids (Table 2), and over 40 isolates have been sequenced. Viroids infect monocots and dicots, and ornamental as well as crop plants.

Classification schemes devised for viroids are based on overall sequence homology (Puchta et al., 1988) or based on homologies in conserved core sequences (Koltunow and Rezaian, 1989). The three groups of viroids are the potato spindle tuber viroid (PSTVd) group, the apple scar skin viroid

TABLE 1
Plant Virus Groups[a]

Characterization	Group or family
Enveloped ssRNA	Rhabdoviridae
	Bunyaviridae: Tospovirus
Nonenveloped ssRNA	
Monopartite genomes	
Isometric particles	Carmovirus
	Luteovirus
	Maize chlorotic dwarf virus group
	Marafivirus
	Necrovirus
	Parsnip yellow fleck virus group
	Sobemovirus
	Tombusvirus
	Tymovirus
Rod-shaped particles	Capillovirus
	Carlavirus
	Closterovirus
	Potexvirus
	Potyvirus
	Tobamovirus
Bipartite genomes	
Isometric particles	Comovirus
	Dianthovirus
	Fabavirus
	Nepovirus
	Pea enation mosaic virus group
Rod-shaped particles	Tobravirus
Tripartite genomes	
Isometric particles	Bromovirus
	Cucumovirus
Isometric and bacilliform particles	Alfalfa mosaic virus group
	Ilarvirus
Rod-shaped particles	Hordeivirus
Quadrapartite genomes	Furovirus
Rod-shaped particles	Tenuivirus
Nonenveloped dsRNA	Reoviridae: Fijivirus, Phytoreovirus
	Cryptovirus
Nonenveloped dsDNA	Badnavirus
	Caulimovirus
	Commelina yellow mottle virus group
Non enveloped ssDNA	Geminivirus

[a] Information adapted from Martelli (1992) and Matthews (1991).

TABLE 2
Diseases Known to Be Incited by Viroids

Disease	Reference
Apple scar skin	Koganezawa *et al.* (1985)
Avocado sunblotch	Thomas and Mohamed (1979)
Burdock stunt	Chen *et al.* (1983)
Citrus exocortis	Semancik and Weathers (1972)
Coconut cadang-cadang	Randles (1975)
Coleus viroid	Singh and Boucher (1991)
Columnea latent	Owens *et al.* (1978)
Chrysanthemum chlorotic mottle	Romaine and Horst (1975)
Cucumber pale fruit	Van Dorst and Peters (1974)
Chrysanthemum stunt	Diener and Lawson (1973)
Coconut tinangaja	Keese *et al.* (1988)
Dapple apple	Hadidi *et al.* (1990)
Grapevine yellow speckle	Koltunow and Rezaian (1988)
Hop stunt	Sasaki and Shikata (1977)
Hop latent	Puchta *et al.* (1988)
Indian bunchy top of tomato	Mishra *et al.* (1991)
Potato spindle tuber	Diener (1971)
Tomato apical stunt	Walter (1981)
Tomato plant macho	Galino *et al.* (1982)

(ASSVd) group, and the avocado sunblotch viroid (ASBVd) group. In contrast to plant viruses, the methods used for the preservation of various viroids are very similar.

Agricultural and Industrial Importance

Many plant viral diseases are of great economic importance, with losses in crop yield and quality in large-acreage crops estimated at $1.5 to $2 billion annually in the United States (Bialy and Klausner, 1986). Viruses infect most crop species, including forest trees (Nienhaus and Castello, 1989), as well as noncrop species, in which they may act as a reservoir for future infections. There is a wide variation in the extent of particular crop losses—losses that are influenced not only by the virus infection, but also by seasonal and regional variations.

Viroids can also cause great problems in a few crops, such as PSTVd in potato, cadang-cadang viroid (CCCVd) in coconut, and chrysanthemum stunt viroid (CSVd) in chrysanthemum, yet can happily survive in other hosts without producing noticeable symptoms (latent infections). In some cases, although the viroid may be latent in one host, it may cause severe symptoms when transferred to another host. One example of this is *Columnea* latent viroid (CLVd), which latently infects *Columnea erythrophae*, an ornamental houseplant, yet causes severe stunting and epinasty in potato and tomato (Owens *et al.*, 1978).

Apart from their economic impact in agriculture by their contribution to crop losses, plant viruses are also growing in value (1) as a source of efficiently expressed promoters (e.g., CaMV 35S promoter) for expression of foreign genes in transgenic plants, (2) as a means of introducing resistance to viruses in crop plants (Beachy *et al.*, 1990), and (3) as a novel means of producing marketable products in plants, e.g., drugs and industrial enzymes (Brisson *et al.*, 1984; De Zoeten *et al.*, 1989; Moffatt, 1991). Therefore, interest in maintaining pure cultures of virus stocks is increasingly of interest to industrial as well as academic laboratories.

Transmission

The natural transmission of viroids from plant to plant is by mechanical means, primarily by plant contact or contaminated tools (Diener, 1987). Many viroids that are not readily mechanically transmissible by rubbing of leaf tissues can be transmitted by grafting of infected buds, scions, and roots [e.g., citrus exocortis viroid (CEVd), and ASBVd]. In addition, some viroids, e.g., PSTVd and *Coleus* viroid, are transmitted through the seed and/or pollen of infected plants (Fernow *et al.*, 1970; Singh, 1970; Singh and Boucher, 1991).

Most plant viruses are mechanically transmissible experimentally even though they may be vector transmitted in the field. In mechanical transmission, wounding or abrading the leaf tissue during inoculation facilitates the entry of the virus inoculum into the plant cell. The type of buffer, its pH, and additives to the buffer will influence the stability of the virus in solution and thereby influence the effectiveness of transmission (Kado, 1972). The effectiveness of transmission is also influenced by the type of virus being transmitted as well as the tissue from which it is extracted and to which it is inoculated. Biological methods for virus transmission include inoculation by dodder, grafting, and vector (insect, fungus, nematode, mite) (Matthews, 1991). Many plant viruses are also seed and pollen transmitted.

Transmission by insects can be defined as nonpersistent, semipersistent, or persistent. Persistent transmission requires that the virus cross the gut wall and enter the haemocael and then the salivary gland before transmission to a new host plant—this period of time is defined as the latent period.

In addition, the virus may be propagative or nonpropagative in the insect; that is, it may or may not replicate in the insect prior to transmission. For example, a propagative leafhopper-borne virus requires multiplication before the insect can transmit the virus.

Aphids constitute the largest group of known plant virus vectors in the field (Watson, 1972). Examples of nonpersistent viruses that are transmitted by aphids include members of the potyvirus and cucumovirus groups. These viruses can also be easily transmitted by mechanical inoculation. Some semipersistent (citrus tristeza virus) (Raccah et al., 1976) and persistent viruses can be mechanically transmitted with difficulty. In these cases, the inoculation of stored virus into the host plant is facilitated by artificial feeding or direct injection of the virus into the aphid in sufficient amounts to result in infection of the plant when the aphid feeds on it [see Watson (1972) for more specifics]. Persistent viruses are those that remain in their insect vector and may replicate (propagative) or not replicate (nonpropagative) in the vector. For example, several members of the Rhabdoviridae family replicate in their aphid vector, whereas members of the luteovirus group do not. Field transmission of these viruses in either case requires the insect vector.

Leafhoppers and planthoppers also account for a significant amount of virus transmission in the field. No viruses have been found to be trasmitted in a nonpersistent manner by these insects. Therefore, transmission of the virus requires a latent period in the insect before transmission can occur. In these cases as well, feeding of the insect on infected plant material or on artificial medium containing the virus, or direct injection of the virus into the insect, is used to facilitate transmission of the virus.

Fungal and nematode vectors, as well as mites, are efficient transmitters of some specific plant viruses in the field. The reader is referred to several literature citations for more information (Matthews, 1991; Taylor, 1972; Teakle, 1972).

Polson and von Wechmar (1980) reported the electrophoretic inoculation of two viruses directly into intact leaves. In their study, brome mosaic virus, a bromovirus, was inoculated into intact leaves of bean, maize streak virus, a leafhopper-transmitted geminivirus, was inoculated into intact leaf tissue of maize. In both instances, electroinfection resulted in the successful establishment of a systemic infection. A second, exciting development in the improvement of the transmission of vector-transmitted viruses has been reported by Brown and Ryan (1991). High-velocity microprojectile particle bombardment of intact host plants using tungsten particles coated with purified virions or partially purified nucleic acids of the whitefly-transmitted watermelon curly mottle isolate of squash leaf curl virus or the chino del tomate virus resulted in establishment of infection.

Characterization

VIROIDS

A common approach to the characterization of most viroids and one that has been used for many years relies on their ready transmissibility by mechanical means. Many of the known viroids can be transmitted to herbaceous hosts, e.g., tomato and cucumber. Unfortunately, most viroids give similar types of symptoms, if any, on these hosts. In addition, most of the symptoms observed with viral diseases also occur with one or more viroid diseases. Therefore, the sole use of indicator hosts is not sufficient to determine if a plant is virus or viroid infected, much less the identity of the viroid. Alternate and more accurate means of characterization can be used to identify viroids from any number of hosts and include bidirectional electrophoresis of nucleic acid preparations from infected tissue (Schumacher *et al.*, 1986; Singh and Boucher, 1987) or molecular hybridization with viroid-specific probes (Owens and Diener, 1981).

Purifications of viroids are based on schemes for the isolation and purification of nucleic acids (Diener *et al.*, 1977). Due to their characteristic mobility under denaturing electrophoretic conditions, further purification from nucleic acid preparations can be made by the gel systems described above. Once pure, the viroid can be sequenced directly (Gross *et al.*, 1978) or cloned using various cDNA cloning techniques (Cress *et al.*, 1983) followed by sequence analysis of the DNAs. If homology to known viroids can be established, polymerase chain reaction using viroid-specific oligonucleotide primers can be used to amplify the genome for sequence analysis and cDNA cloning (Hadidi and Yang, 1990). Once the sequence of the viroid is known, it is then possible to compare that sequence with known viroids to determine to which group it belongs.

VIRUSES

Plant viruses are characterized by a number of criteria, including the production of characteristic symptoms on host indicator plants, and a number of physical and biochemical techniques, such as serology, electron microscopy, and nucleic acid hybridization.

Some viruses have very narrow host ranges, limited to a single genus, whereas others have a broad host range, such as cucumber mosaic virus (Francki *et al.*, 1979) and tobacco mosaic virus (Zaitlin and Israel, 1975). Even within a single virus group, there may be significant differences. Most potyviruses, for example, have narrow host ranges; garlic yellow stripe virus, for example, is only known to infect *Allium sativum* and *Chenopodium murale* (Mohamed and Young, 1981). However, bean yellow mosaic virus infects many legumes, including black locust and a number of bulbous ornamentals (Bos, 1970), and has even been reported to infect beech trees (Winter and Nienhaus, 1989). For identification and characterization of viruses, studies are often restricted to a relatively narrow range of herbaceous host plants that can be conveniently grown in greenhouses. Among the biological characteristics that are commonly examined are the host range and types of symptoms produced. Both of these criteria are rather variable, with significant variation possible between genotypes of the same plant species (Van der Want *et al.*, 1975) and under different environmental conditions. If the host range of a "new" isolate is to be compared to published data, it is desirable to retest a type isolate under identical conditions for comparison. In all cases, the source tissue should be from an active infection, because older tissue may have much lower levels of infectious virus. Other characteristics that were once commonly used, but have now been generally replaced (Francki, 1980) because of the variability of source tissue and between host species, are longevity *in vitro* (measured by determining infectivity of sap extracts after defined periods of storage), thermal inactivation point (termperature required to abolish infectivity, usually after 10 min of incubation at the specified temperature), and dilution end point.

Serological techniques can be employed either to discriminate between very closely related entities or to demonstrate relationships between distantly related ones. Various forms of assay, each with advantages and disadvantages, are described by Van Regenmortel (1982). It should be borne in mind that the method used for preparation of the antiserum and length of immunization can significantly affect the specificity of the antiserum, and that different animals of the same species may yield antisera with quite different reactions (Van Regenmortel, 1982).

Electron microscopic examination of negatively stained (Hitchborn and Hills, 1965) sap extracts or partially purified virus preparations can yield much information, which may be sufficient in some cases to assign a virus to a specific group. Electron microscopy is also an excellent way of detecting contaminants and mixed infections. Ultrastructural examination of embedded material may also be valuable, especially for an experienced operator; a novice may misinterpret variations due to physiological state or other pathologies as effects of the virus.

Electrophoretic analyses of the structural proteins and nucleic acids are also used to characterize plant viruses. Analysis of nonstructural proteins (Hiebert and McDonald, 1973) and double-stranded RNAs (Valverde *et al.*, 1986) involved in replication is increasingly finding favor, both for comparing related viruses and for discriminating between strains.

Nucleic acid hybridization (Rosner and Bar-Joseph, 1984) and related techniques such as the polymerase chain reaction (Langeveld *et al.*, 1991; Vunsh *et al.*, 1990) are also becoming more common for the detection and identification of plant viruses, and also for strain discrimination (Gould and Symons, 1983).

Purification   Among the factors to be considered (e.g., Matthews, 1991) are the type of host plant used for virus propagation, virus stability, and the use for which the purified preparation is intended. In many cases, the most highly purified preparation will have significantly lower specific infectivity than it is possible to recover in a slightly less pure form.

Some viruses are highly stable, and tissue can be frozen and accumulated prior to purification. This can have the advantage of causing aggregation of some host proteins, aiding in their separation from the virus during purification. Potyviruses, among others, are more likely to be damaged by freezing, resulting in significantly lower yields than can be obtained from fresh tissue. Antioxidants and additives such as polyvinylpyrrolidone may also be used to minimize losses due to aggregation and tannins; excessive centrifugal force may also cause irreversible aggregation of some viruses. Salting out or acidification may be used successfully with some viruses to cause precipitation of host components, leaving virus in the supernatant. Precipitation of virus with polyethylene glycol (Hebert, 1963) is also commonly used to separate virus from host components as part of a series of differential centrifugations. However, after any type of precipitation it is important to resuspend the virus fully before the next step of purification. Rate zonal sucrose density gradients (Brakke, 1960) are commonly used to separate virus particles from other components of the plant extract— especially fraction 1 protein and phytoferritin. An alternative is quasi-equilibirum isopycnic gradients of cesium chloride (Meselson *et al.*, 1957) or cesium sulfate, but these may offer improved purity at the cost of specific infectivity. If the virus is intended as antigen to raise antiserum, the loss of infectivity is less important than the increase in purity, but absolute purity may not always be as critical as infectivity.

Methods for Preservation

Plant virus or viroid collections stored in stock plants and passaged to maintain infections have the disadvantages of risk of contamination and loss of cultures, mutation of the original isolate through passaging, and the large amount of space needed to maintain plants.

The methods that have proved most successful in the preservation of viroids and viruses have been those involving dehydration of infected tissue (McKinney, 1953) and lyophilization of infected sap. Ultralow-temperature freezing of fresh tissue and storage of purified virus are also effective for many viruses. A critical component of plant virus preservation is the ability to transmit the stored virus back to a host plant. This is especially important in the case of vector-transmitted viruses. Culture of infected tissues is an alternative, although not as practical or practiced as the use of lyophilization. A more recent alternative to the preservation of distinct strains of viruses and viroids is the use of cloned, infectious molecules. Each of these methods of preservation are discussed below.

Dehydration of Infected Tissue

A simple, but effective means of preserving many plant viruses and viroids is to dehydrate infected tissue over anhydrous $CaCl_2$, followed by storage of sealed bottles or ampules at 4°C or frozen at −20 or at −70°C. Incomplete drying of samples in the humid atmosphere of a laboratory may result in the loss of stability/infectivity of the virus (McKinney, 1947). In addition to stability for infectivity, drying over $CaCl_2$ is suitable for serology (Armitage et al., 1989) or electron microscopy (Bos and Benetti, 1979). Table 3 summarizes successful techniques used in the dehydration of leaf tissue infected with plant viruses representing several virus groups as well as viroids. Briefly, fresh virus- or viroid-infected leaves are collected and processed as quickly as possible. [Note: The choice of which tissue to collect will depend on the titer of the pathogen in that particular host tissue.] With large leaves, the midrib is removed and the remainder of the tissue (leaf or otherwise) is cut into strips about 2 cm in width. The leaf samples can be (1) desiccated over $CaCl_2$ or silica gel at room temperature or at 4°C and then stored in sealed dark bottles or ampules in a desiccator at room temperature, over $CaCl_2$ at 1–2°C, or over 1% magnesium perchlorate (Horvath and Besada, 1980; McKinney, 1945, 1947; McKinney et al., 1965; McKinney and Silber, 1968; Pocsai, 1986; Zaumeyer, 1962), or at −20°C (Kirkpatrick, 1962), or (2) frozen quickly in liquid nitrogen and stored at −20 or −70°C (McKinney et al., 1961; Best, 1961).

An important observation made by McKinney et al. (1965) was that host plants differ in their value for virus storage, e.g., for barley stripe mosaic virus, barley and wheat were superior to oats for storage of the virus. In addition, Horvath and Besada (1980) stress that the propagative host plays an important role in determining the activity of a virus after a certain period of preservation. It has also been observed for tomato spotted wilt virus, a tospovirus, that is important to colelct tissue to be used for preservation before any tissue on the plant turns necrotic from virus infection. Once any tissue has turned necrotic the virus present in nonnecrotic tissue loses infectivity more rapidly (J. Moyer, personal communication).

Cryopreservation

A widely adopted method for the preservation of plant viruses is the storage of infected tissue or partially purified virus preparations that have

TABLE 3

Examples of Published Successful Techniques for Preservation of Plant Viruses and Viroids by Dehydration and Cryopreservation

Virus group	Virus	Method of preservation	Additive	Storage	Known longevity[a]	Reference
Dehydration of leaf tissue						
Ilarvirus	Prunus necrotic ringspot	Silica gel	—	−22°C	6 years	Kirkpatrick (1962)
	Prune dwarf	Silica gel	—	−22°C	1 year	Kirkpatrick (1962)
Sobemovirus	Southern bean mosaic	$CaCl_2$	—	4°C	8 years	Zaumeyer (1962)
AlMV	Alfalfa mosaic	$CaCl_2$	—	4°C	7 years	Zaumeyer (1962)
Potyvirus	Bean yellow mosaic	$CaCl_2$	—	4°C	31 months	Zaumeyer (1962)
AlMV	Alfalfa mosaic	$CaCl_2$	—	Over Mg perchlorate	16.5 years	McKinney et al. (1965)
Hordeivirus	Barley stripe mosaic	$CaCl_2$	—	Over Mg perchlorate	16 years	McKinney et al. (1965)
Bromovirus	Brome mosaic	$CaCl_2$	—	Over Mg perchlorate	16 years	McKinney et al. (1965)
Cucumovirus	Cucumber mosaic	$CaCl_2$	—	Over Mg perchlorate	19 years	McKinney et al. (1965)
Potyvirus	Tobacco etch	$CaCl_2$	—	Over Mg perchlorate	17 years	McKinney et al. (1965)
Ilarvirus	Tobacco streak	$CaCl_2$	—	Over Mg perchlorate	8 years	McKinney et al. (1965)
Potexvirus	Potato virus X	$CaCl_2$	—	Over Mg perchlorate	15 years	McKinney et al. (1965)
Furovirus	Wheat mosaic (soilborne)	$CaCl_2$	—	Over Mg perchlorate	11 years	McKinney et al. (1965)
24 isolates of 19 viruses	—	$CaCl_2$	—	4°C	10–87 months	Horvath and Besada (1980)
Tobamovirus	Tobacco mosaic	Dried	—	RT[b]	50 years	Silber and Burk (1965)
Cryopreservation						
Tospovirus	Tomato spotted wilt	Frozen tissue	—	−70°C	6 years	Best (1961)
Luteovirus	Barley yellow dwarf	Sap	Albumin	−12–4°C	12 months	Gill (1973)
		Virus	Glycerol/sucrose	Liquid N/lyophilization	4 years	Rochow et al. (1976)
Sobemovirus	Southern bean mosaic	Sap	Lactose	Frozen	2 years	Worley and Schneider (1966)
		Virus	Lactose	Lyophilization	8 years	Worley and Scheider (1966)
46 isolates of 39 viruses		Sap	Peptone/glucose	Lyophilization RT	4–11 months (5 years for some)	Hollings and Lelliott (1960)
74 viruses		Sap	Peptone/glucose	RT	1–10 years	Hollings and Stone (1970)
Potato spindle tuber viroid		Frozen, dried tissue	—	−70°C	6 years	Singh and Finnie (1977)
		Lyophilized nucleic acid	—	−70°C	2 years	Singh and Bagnall (1968)

[a] These values do not necessarily represent maximum attainable storage times, nor do they reflect differences encountered from different host tissues, but should serve as a guide for selection of preservation methods. [b] RT, Room temperature.

been preserved by freezing or freeze-drying and stored in sealed ampules or tubes at low temperatures. The American Type Culture Collection (ATCC) (1991) suggests that most mechanically transmitted plant viruses can be successfully stored by freezing and freeze-drying.

FREEZING METHODS

Preservation by freezing and storage at low temperature has long been used for the storage of plant viruses (DeWijs and Suda-Bachmann, 1979; McKinney *et al.*, 1961; Rochow *et al.*, 1976). The maintenance of infectivity of plant viruses frozen directly in leaf tissue benefits from the use of ultralow temperatures for both the freezing and storage of the material (Best, 1961). Freezing tissue at $-70°C$ is used for the storage of many of the more stable viruses. In addition, clarified sap or purified virus preparations may be stored for long periods of time at $-20°C$ or in liquid nitrogen ($-196°C$). Fresh tissue containing potyviruses has retained infectivity when stored frozen at $-70°C$ for several years (J. Hammond, unpublished). It is important for some viruses that the cooling rate be carefully controlled during freezing (e.g., tobacco streak virus; L. McDaniel, personal communication). The effect of freezing on the structure of turnip yellow mosaic virus was examined in a systematic study by Kaper and Siberg (1969a,b). Freezing in water or in 0.1 M sodium acetate pH 6.0 at a virus concentration of 0.3–0.5% at $-75°C$ was followed by thawing at room temperature. When the virus was frozen in water, only a very small amount of virus was recovered intact; the 0.1 M sodium acetate solution, on the other hand, afforded some protection of the stability of the virus. Other additives, such as 10% glycerol (Rochow *et al.*, 1976, 10% lactose (Worley and Schneider, 1966), or 7% glucose plus 7% peptone (Hollings and Lelliott, 1960) have been successfully used as cryoprotectants for clarified sap and virus concentrates. Alternatively, isolates of tobacco mosaic virus, brome mosaic virus, and cowpea chlorotic virus have been stored in a buffer containing 5% ethylene glycol (A. O. Jackson, personal communication). In addition, for a number of viruses, leaf tissue stored at $-20°C$ in glycerol or vacuum-infiltrated with glycerol and stored frozen retains infectivity for at least 2 years.

The most consistent method for the preservation of ilarviruses, e.g., prunus necrotic ringspot virus, has been the storage of hand-collected pollen from the flowers of infected trees (J. Crosslin and G. Mink, personal communication). Virus present in pollen stored frozen or at 4°C has retained its infectivity for up to 6 years (from the time of initial storage to the present). Most ilarviruses have not survived more than 2 years in storage in dried leaf tissue. It is also important that hand-collected pollen and not bee-stored pollen be collected, as there is an agent in the bee-stored pollen that degrades the virions of the virus (Cole and Mink, 1984).

Viruses of the family Rhabdoviridae can be extracted and stored at $-70°C$ in small aliquots in a buffer containing 100 mM Tris, 10 mM magnesium acetate, 40 mM sodium sulfite, and 1 mM $MnCl_2$ (A. O. Jackson, personal communication; Jackson *et al.*, 1987). Preservation of viable potato yellow dwarf virus was accomplished when purified virus in a 0.1 M glycine, 0.01 M $MgCl_2$, pH 7.0 solution was stored at $-80°C$ or lower (H.-T. Hsu, personal communication). Assays in vector cell monolayers have not yet revealed loss of infectivity since the time the onoculum was prepared and frozen in 1969. Long-term preservation of infectivity was also attained for wound tumor virus, a phytoreovirus, when inocula were prepared in 0.1 M histidine, 0.01 M $MgCl_2$, pH 6.4 and stored at $-80°C$ (H.-T. Hsu, personal communication).

Purified viroid preparations and gel-purified viroids have retained infectivity for many years when stored in small aliquots at -20 to $-70°C$. The addition of cryoprotectants does not seem to be necessary for maintenance of viroid stability.

FREEZE-DRYING

Plant viruses of many taxonomic groups can be stored as freeze-dried tissue, clarified sap, or purified virus preparations (Table 3). The ATCC (L. McDaniel, personal communication) suggests that leaf pieces be placed in vials and gradually cooled (e.g., $-1°C/min$) in either a computer-controlled freezing unit or, more inexpensively, in a -20 or $-70°C$ freezer (no defrost cycle) for several hours using small, isopropanol-filled containers now commercially available for controlled freezing. If necessary, containers of various sizes may also be constructed from common laboratory materials and pretested for rate of freezing or efficacy in maintenance of virus infectivity. The samples are then freeze-dried under vacuum and stored at $-70°C$. These virus samples can be reconstituted by grinding the tissues in 0.05 M sodium phosphate and 10 mM sodium sulfite, pH 7.2, followed by standard inoculation for mechanically transmitted viruses (see ATCC, 1991).

Tissue Culture

Timoshenko *et al.* (1989) describe conditions used for *in vitro* cultivation and maintenance of potato virus Y in *Nicotiana glutinosa* callus cultures with the possibility of long-term storage. Briefly, 1 month after inoculation, leaves were taken, surface sterilized, and put into culture to induce the formation of callus. The cultures were examined by electron microscopy and serology and were found to contain the virus. The authors suggest that rapid growth of the callus may lead to virus elimination, so care should be taken to ensure that the virus maintains its replication ability, etc. This method of virus preservation has been proposed for virus-infected woody tissue (apples, prunus). Callus or cell cultures may be maintained at low temperatures (4°C) for long-term maintenance, reducing the necessity for transfers of cultures. [For more information on plant tissue culture maintenance, see Withers (1991).]

An alternate method for the preservation of vector-transmitted, propagative viruses may be the use of infected insect cell monolayers. Kimura and Black (1972) showed that synchronous infection of wound tumor virus could be achieved with ease in AC20 monolayer cells of *Agallia constricta* leafhopper. Hsu and Black (1974) infected a monolayer cell culture of AS2 (*Aceratagallia sanguinolenta*), the leafhopper vector of sanguinolenta potato yellow dwarf virus, a member of the Rhabdoviridae. Cells were inoculated with a purified virus suspension and 100% infection was achieved within 9 hours postinoculation. The concentration of progeny virus was maintained at a plateau level for 52 hours postinoculation, at which time the experiments were ended. It may, therefore, be possible that infected insect cell lines could be used as a method for virus storage and maintenance.

Mülbach and Sänger (1981) reported the continuous replication of potato spindle tuber viroid (PSTVd) in callus cultures from PSTVd-infected wild-type potato and tomato plants and in cell suspensions derived from potato protoplasts. The persistence of viroid replication was monitored through 14 subculture passages, corresponding to continuous replication over a period of more than 1 year. It is feasible, therefore, that long-term maintenance of viroid-infected plant material can be performed through tissue culture. A question arises, however, of the stability of the nucleotide sequence through many passages.

Recombinant Molecules

The use of cloned, infectious copies of plant viruses and viroids has revolutionized the study of genome expression strategies and functional determination of various viral genes and viroid structure/function relationships. It is also a means of preservation of specific isolates of these pathogens. Drawbacks of this method include the technical difficulties encountered in and expertise required for the cloning full-length copies of the genomic nucleic acids, as well as the production of infectious molecules from these clones. Nevertheless, this strategy has proved quite successful in many cases, and the list of cloned, infectious viruses and viroids is steadily growing (Table 4; review of infectious cDNA clones of plant RNA viruses, Boyer and Haenni, 1994).

Evaluation of Viability and Stability

The methods described above provide information on preservation; however, continual monitoring of viability levels, genetic stability, and

TABLE 4
Examples of Cloned, Infectious Viroids

Viroid	Reference
Citrus exocortis viroid	Visvader *et al.* (1985); Rigden and Rezaian (1992)
Hop stunt viroid	Ohno *et al.* (1983); Shikata *et al.* (1984); Meshi *et al.*, (1985)
Potato spindle tuber viroid	Cress *et al.* (1983); Tabler and Sänger (1984)
Tomato apical stunt viroid	Owens (1990); Candresse *et al.* (1987)
Columnea latent viroid	Hammond *et al.* (1989)

purity must be conducted. The method used for the evaluation of viability depends on the method used for preservation and the mode of transmission of the virus or viroid.

Reconstitution of plant viruses and viroids that have been stored as dehydrated tissue or as lyophilized samples for the purpose of inoculation onto host plants can be performed by triturating the material in a buffer, such as 0.05 M sodium phosphate, pH 7.2, containing 10 mM sodium bisulfite (ATCC, 1991, p. 22). If samples are frozen sap, they are thawed at room temperature, or 37°C. The suspensions can be used to inoculate mechanically the leaves of host plants when used in combination with an abrasive such as Celite (diatomaceous earth) or carborundum (silicon carbide) (Matthews, 1991).

Effective inoculation of vector-transmitted viruses requires inoculation of the insects by feeding or by direct injection of the virus into the insect. In a study of the stability of barley yellow dwarf virus (BYDV), an aphid-transmitted, phloem-limited virus, Gill (1973) reconstituted lyophilized virus with water. Alternatively, frozen sap samples were thawed. Membrane-feedine of insects was performed by the method of Rochow and Brakke (1964), or by injecting diluted virus into the aphids. Unfortunately, long-term culture of vector-transmitted viruses in plants without going through the insect can result in loss of the ability of the virus to be transferred (Matthews, 1991). ATCC recommends that viruses that are mechanically transmissible with difficulty or transmissible only by insect vectors be obtained from fresh tissues from investigators actively working with these viruses.

Transmission of cloned virus or viroid molecules ranges from direct inoculation of cloned DNA copies of the virus or viroid to leaf tissue,

TABLE 5
Examples of Cloned, Infectious Plant Viruses

Group	Virus	Reference
DNA		
Badnavirus	Rice tungro bacilliform virus	Dasgupta *et al.* (1991)
Caulimovirus	Cauliflower mosaic virus	Howell *et al.* (1980); Lebeurier *et al.* (1980)
Geminivirus	Maize streak virus	Lazarowitz *et al.* (1989)
	Cassava latent virus	Stanley (1983)
	Tomato yellow leaf curl virus	Rochester *et al.* (1990); Navot *et al.* (1991)
	Squash leaf curl virus	Lazarowitz and Lazdins (1991)
RNA		
Bromovirus	Brome mosaic virus	Ahlquist *et al.* (1984); Janda *et al.* (1987); Pogany *et al.* (1994)
	Cowpea chlorotic mottle virus	Allison *et al.* (1988)
Carmovirus	Turnip crinkle virus	Collmer *et al.* (1992); Heaton *et al.* (1989)
Comovirus	Cowpea mosaic virus	Holness *et al.* (1989); Dessens and Lomonossoff (1993)
	Cowpea severe mosaic virus	Chen and Bruening (1992)
Cucumovirus	Cucumber mosaic virus	Rizzo and Palukaitis (1990)
Dianthovirus	Red clover necrotic mosaic virus	Xiong and Lommel (1991)
Furovirus	Beet necrotic yellow vein virus	Jupin *et al.* (1992); Ziegler-Graff *et al.* (1988)
Hordeivirus	Barley stripe mosaic virus	Petty *et al.* (1989)
Luteovirus	Barley yellow dwarf virus	Young *et al.* (1991)
Potexvirus	Potato virus X	Hemenway *et al.* (1990)
	White clover mosaic virus	Beck *et al.* (1990)
Potyvirus	Plum pox virus	Riechmann *et al.* (1990); Maiss *et al.* (1992)
	Tobacco vein mottling virus	Domier *et al.* (1989)
Tobamovirus	Tobacco mosaic virus	Meshi *et al.* (1986), Dawson *et al.* (1986)
Tobravirus	Pea early browning virus	MacFarlane *et al.* (1991; 1992)
	Cucumber necrosis virus	Rochon and Johnson (1991)
	Cymbidium ringspot virus	Burgyan *et al.* (1990)
Tombusvirus	Tomato bushy stunt virus	Hearne *et al.* (1990)
Tymovirus	Turnip yellow mosaic virus	Weiland and Dreher (1989)

to preparation of infectious transcripts and inoculation of the transcripts onto host plants—in either case a phosphate buffer containing an abrasive may be added to facilitate transmission. Improvements in infectivity of cloned viruses have been reported, most of which involve additional cloning procedures, e.g., the use of the cauliflower mosaic virus transcription promoter (CaMV 35S) to increase specific infectivity, e.g., pea early browning virus (MacFarlane et al., 1992), plum pox virus (Maiss et al., 1992), and brome mosaic virus (Mori et al., 1991). Additional nucleotides, primarily at the 5' terminus of in vitro transcripts, may reduce infectivity, e.g., in cowpea mosaic virus (Eggen et al., 1989), brome mosaic virus (Janda et al., 1987), and satellite tobacco mosaic virus (Mirkov et al., 1990). In other cases, however, the presence of one or two additional nucleotides at the 5' end did not affect infectivity, (e.g., in cucumber mosaic virus (Rizzo and Palukaitis, 1990) and cowpea mosaic virus (Vos et al., 1988).

Mechanical inoculation of infectious cloned DNA of the geminivirus, cassava latent virus, was reported by Ward et al. (1988). Rubbing native or cloned CaMV onto leaves using a buffer and abrasive powder will also result in infection (Shepherd et al., 1970; Lebeurier et al., 1980). In addition, Agrobacterium-mediated inoculation has been used successfully to transmit both viroid (Gardner et al., 1986) and vector-transmitted geminiviruses (Lazarowitz et al., 1989; Lazarowitz and Lazdins, 1991; Rochester et al., 1990).

Investigation of the genetic stability of infectious DNAs and transcripts of cloned viruses and viroids has revealed that sequence heterogeneity in the progeny can occur with some frequency when plants are inoculated with transcripts of the satellite of cucumber mosaic virus (Collmer and Kaper, 1988; Kurath and Palukaitis, 1990) and transcripts of tomato apical stunt viroid (TASVd) (Owens, 1990).

Summary

We hope that this chapter will provide an informative starting point for those novices wishing to use various preservation techniques for their respective virus or viroid sample. Again, we refer the reader to the references cited for more specific information regarding their particular situation. Plant viruses and viroids and molecularly cloned plant viruses and viroids can be obtained from the American Type Culture Collection, Virus Collection Manager, 12301 Parklawn Drive, Rockville, Maryland 20852.

Acknowledgments

The authors would like to thank Dr. L. McDaniel of the American Type Culture Collection for providing information used in the preparation of this chapter and Dr. L. McDaniel and Dr. Hei-Ti Hsu for critical reading of the manuscript.

References

Ahlquist, P., French, R., Janda, M., and Loesch-Fries, L. S. (1984). *Proc. Natl. Acad. Sci. U.S.A.* **81,** 7066–7070.

Allison, R. F., Janda, M., and Ahlquist, P. (1988). *J. Virol.* **62,** 3581–3588.

American Type Culture Collection (ATCC) (1991). "ATCC Preservation Methods: Freezing and Freeze-drying," 2nd ed., p. 22. ATCC, Rockville, MD.

Armitage, C. R., Hunger, R. M., Sherwood, J. L., and Weeks, D. L. (1989). *J. Phytopathol.* **127,** 116–122.

Beachy, R. N., Loesch-Fries, S., and Tumer, N. E. (1990). *Annu. Rev. Phytopathol.* **28,** 451–474.

Beck, D. L., Forster, R. L. S., Bevan, M. W., Boxen, K. A., and Lowe, S. (1990). *Virology* **177,** 152–158.

Best, R. J. (1961). *Virology* **14,** 440–443.

Bialy, H., and Klausner, A. (1986). *Bio/Technology* **4,** 96.

Bos, L. (1970). *CMI/AAB Description Plant Viruses* No. 40.

Bos, L., and Benetti, M. P. (1979). *Neth. J. Plant. Pathol.* **85,** 241–251.

Boyer, J.-C., and Haenni, A.-L. (1994). *Virology* **198,** 415–426.

Brakke, M. K. (1960). *Adv. Virus Res.* **7,** 193–224.

Brisson, N., Paszkowski, J., Penswick, J. R., Gronenbron, B., Potrykus, I., and Hohn, T. (1984). *Nature (London)* **310,** 511–514.

Brown, J. K., and Ryan, R. (1991). *Phytopathology* **81,** 1217.

Burgyan, J., Nagy, P. D., and Russo, M. (1990). *J. Gen. Virol.* **71,** 1857–1860.

Candresse, T., Smith, D., and Diener, T. O. (1987). *Nucleic Acids Res.* **15,** 597.

Chen, W., Tien, P., Zhu, Y. X., and Liu, Y. (1983). *J. Gen. Virol.* **64,** 409–414.

Chen, X., and Bruening, G. (1992). *Virology* **191,** 607–618.

Cole, A., and Mink, G. I. (1984). *Phytopathology* **74,** 1320–1324.

Collmer, C. W., and Kaper, J. M. (1988). *Virology* **163,** 293–298.

Collmer, C. W., Stenzler, L., Chen, X., Fay, N., Hacker, D., and Howell, S. H. (1992). *Proc. Natl. Acad. Sci. U.S.A.* **89,** 309–313.

Cress, D. E., Kiefer, M. C., and Owens, R. A. (1983). *Nucleic Acids Res.* **11,** 6821–6835.

Dasgupta, I., Hull, R., Eastop, S., Poggi-Pollini, C., Blakebrough, M., Boulton, M. I., and Davies, J. W. (1991). *J. Gen. Virol.* **72,** 1215–1221.

Dawson, W. O., Beck, D. L., Knorr, D. A., and Grantham, G. L. (1986). *Proc. Natl. Acad. Sci. U.S.A.* **83,** 1832–1836.

Dessens, J. T., and Lomonossoff, G. P. (1993). *J. Gen. Virol.* **74,** 889–892.

DeWijs, J. J., and Suda-Bachmann, F. (1979). *Neth. J. Plant. Pathol.* **85,** 23–29.

De Zoeten, G. A., Penswick, J. R., Horisberger, M. A., Ahl, P., Schuyltze, M., and Hohn, T. (1989). *Virology* **172,** 213–222.

Diener, T. O. (1971). *Virology* **45,** 411–428.

Diener, T. O., ed. (1987). "The Viroids." Plenum, New York.

Diener, T. O., and Lawson, R. H. (1973). *Virology* **51,** 94–101.

Diener, T. O., Hadidi, A., and Owens, R. A. (1977). *Methods Virol.* **6,** 185–217.

Domier, L. L., Franklin, K. M., Hunt, A. G., Rhoads, R. E., and Shaw, J. G. (1989). *Proc. Natl. Acad. Sci. U.S.A.* **86**, 3509–3513.

Eggen, R., Verver, J., Wellink, J., de Jong, A., Goldbach, R., and van Kammen, A. (1989). *Virology* **173**, 447–455.

Fernow, K. H., Peterson, L. C., and Plaisted, R. L. (1970). *Am. Potato J.* **47**, 75–80.

Francki, R. I. B. (1980). *Intervirology* **13**, 91–98.

Francki, R. I. B., Mossop, D. W., and Hatta, T. (1979). *CMI/AAB Description Plant Viruses* No. 213.

Galindo, J. A., Smith, D. R., and Diener, T. O. (1982). *Phytopathology* **72**, 49–54.

Gardner, R. C., Chonoles, K., and Owens, R. A. (1986). *Plant Mol. Biol.* **6**, 221–228.

Gill, C. C. (1973). *Plant Dis. Rep.* **57**, 862–864.

Gould, A. R., and Symons, R. H. (1983). *Annu. Rev. Phytopathol.* **21**, 179–199.

Gross, H. J., Domdey, H., Lossow, C., Jank, P., Raba, M., Alberty, H., and Sänger, H. L. (1978). *Nature (London)* **273**, 203–208.

Hadidi, A., and Yang, X. (1990). *J. Virol. Methods* **30**, 261–270.

Hadidi, A., Huang, C., Hammond, R. W., and Hashimoto, J. (1990). *Phytopathology* **80**, 263–268.

Hammond, R. W., Smith, D. R., and Diener, T. O. (1989). *Nucleic Acids Res.* **17**, 10083–10094.

Hearne, P. Q., Knorr, D. A., Hillman, B. I., and Morris, T. J. (1990). *Virology* **177**, 141–151.

Heaton, L. A., Carrington, J. C., and Morris, T. J. (1989). *Virology* **170**, 214–218.

Hebert, T. T. (1963). *Phytopathology* **53**, 362.

Hemenway, C., Weiss, J., O'Connell, K., and Tumer, N. E. (1990). *Virology* **175**, 365–371.

Hiebert, E., and McDonald, J. G. (1973). *Virology* **56**, 349–361.

Hitchborn, J. H., and Hills, G. J. (1965). *Virology* **27**, 528–540.

Hollings, M., and Lelliott, R. A. (1960). *Plant Pathol.* **9**, 63–66.

Hollings, M., and Stone, O. M. (1970). *Ann. Appl. Biol.* **65**, 411–418.

Holness, C. L., Lomonossoff, G. P., Evans, D., and Maule, A. J. (1989). *Virology* **172**, 311–320.

Horvath, J., and Besada, W. H. (1980). *J. Plant Dis. Prot.* **87**, 463–472.

Howell, S. H., Walker, L. L., and Dudley, R. K. (1980). *Science* **208**, 1265–1267.

Hsu, H. T., and Black, L. M. (1974). *Virology* **59**, 331–334.

Jackson, A. O. J., Francki, R. I. B., and Zuidema, D. (1987). *In* "The Rhabdoviruses" (R. R. Wagner, ed.), pp. 427–506. Plenum, New York.

Janda, M., French, R., and Ahlquist, P. (1987). *Virology* **158**, 259–262.

Jupin, I., Guilley, H., Richards, K. E., and Jonard, G. (1992). *EMBO J.* **11**, 479–488.

Kado, C. I. (1972). *In* "Principles and Techniques in Plant Virology" (C. I. Kado and H. O. Agrawal, eds.), pp. 3–31. Van Nostrand-Reinhold, New York.

Kaper, J. M., and Siberg, R. A. (1969a). *Cryobiology* **5**, 366–374.

Kaper, J. M., and Siberg, R. A. (1969b). *Virology* **3**, 407–413.

Keese, P., Osorio-Keese, M. E., and Symons, R. H. (1988). *Virology* **162**, 508–510.

Kimura, I., and Black, L. M. (1972). *Virology* **48**, 852–854.

Kirkpatrick, H. C. (1962). *Plant Dis. Rep.* **46**, 628–630.

Koganezawa, H. (1985). *Ann. Phytopathol. Soc. Jpn.* **51**, 176–182.

Koltunow, A. M., and Rezaian, M. A. (1988). *Nucleic Acids Res.* **16**, 849–864.

Koltunow, A. M., and Rezaian, M. A. (1989). *Intervirology* **30**, 194–201.

Kurath, G., and Palukaitis, P. (1990). *Virology* **176**, 8–15.

Langeveld, S. A., Dore, J.-M., Memelink, J., Derks, A. F. L. M., van der Vlugt, C. I. M., Asjes, C. J., and Bol, J. (1991). *J. Gen. Virol.* **72**, 1531–1541.

Lazarowitz, S. G., and Lazdins, I. B. (1991). *Virology* **180**, 58–69.

Lazarowitz, S. G., Pinder, A. J., Damsteegt, V. D., and Rogers, S. G. (1989). *EMBO J.* **8**, 1023–1032.

Lebeurier, G., Hirth, L., Hohn, T., and Hohn, B. (1980). *Gene* **12**, 139–146.

MacFarlane, S. A., Wallis, C. V., Taylor, S. C., Goulden, M. G., Wood, K. R., and Davies, J. W. (1991). *Virology* **182**, 124–129.

MacFarlane, S. A., Gilmer, D., and Davies, J. W. (1992). *Virology* **187**, 829–831.

Maiss, E., Timpe, U., Brisske-Rode, A., Lesemann, D.-E., and Casper, R. (1992). *J. Gen. Virol.* **73**, 709–713.

Martelli, G. P. (1992). *Plant Dis.* **76**, 436–442.

Matthews, R. E. F. (1991). "Plant Virology," 3rd ed. Academic Press, London.

McDaniel, L. L. (1989). *ATCC Q. Newsl.* **9**(4), 1.

McKinney, H. H. (1945). *Phytopathology* **35**, 488.

McKinney, H. H. (1947). *Phytopathology*, **37**, 139–142.

McKinney, H. H. (1953). *Ann. N. Y. Acad. Sci.* **56**, 615–620.

McKinney, H. H., and Silber, G. (1968). *Methods Virol.* **4**, 491–501.

McKinney, H. H., Greeley, L. W., and Clark, W. A. (1961). *Plant Dis. Rep.* **45**, 755.

McKinney, H. H., Silber, G., and Greeley, L. W. (1965). Phytopathology **55**, 1043–1044.

Meselson, M., Stahl, F. W., and Vinograd, J. (1957). *Proc. Natl. Acad. Sci. U.S.A.* **43**, 581–588.

Meshi, T., Ishikawa, M., Watanabe, Y., Yamaya, J., Okada, Y., Sano, T., and Shikata, E. (1985). *Mol. Gen. Genet.* **200**, 199–206.

Meshi, T., Ishikawa, M., Matoyoshi, F., Semba, K., and Okada, Y. (1986). *Proc. Natl. Acad. Sci. U.S.A.* **83**, 5043–5047.

Mirkov, T. E., Kurath, G., Mathews, D. M., Elliot, K., Dodds, J. A., and Fitzmaurice, L. (1990). *Virology* **179**, 395–402.

Mishra, M. D., Hammond, R. W., Owens, R. A., Smith, D. R., and Diener, T. O. (1991). *J. Gen. Virol.* **72**, 1781–1785.

Moffatt, A. S. (1991). *Genet. Eng. News* **11**(10), 1.

Mohamed, N. A., and Young, B. R. (1981). *Ann. Appl. Biol.* **97**, 65–74.

Mori, M., Mise, K., Kobayashi, K., Okuno, T., and Furusawa, I. (1991). *J. Gen. Virol.* **72**, 243–246.

Mühlbach, H.-P., and Sänger, H. L. (1981). *Biosci. Rep.* **1**, 79–87.

Navot, N., Pichersky, E., Zeidan, M., Zamir, D., and Czosnek, H. (1991). *Virology* **185**, 151–161.

Nienhaus, F., and Castello, J. D. (1989). *Annu. Rev. Phytopathol.* **27**, 165–186.

Ohno, T., Ishikawa, M., Takamatsu, N., Meshi, T., Okada, Y., Sano, T., and Shikata, E. (1983). *Proc. Jpn. Acad., Ser. B* **59**, 251–254.

Owens, R. A. (1990). *Mol. Plant Microbe Interact.* **3**, 374–380.

Owens, R. A., and Diener, T. O. (1981). *Science* **213**, 670–672.

Owens, R. A., Smith, D. R., and Diener, T. O. (1978). *Virology* **89**, 388–394.

Petty, I. T. D., Hunter, B. G., Wei, N., and Jackson, A. O. (1989). *Virology* **171**, 342–349.

Pocsai, E. (1986). *Acta Phytopathol. Entomol. Hung.* **21**, 287–290.

Pogany, J., Huang, Q., Romero, J., Nagy, P. D., and Bujarski, J. J. (1994). *J. Gen. Virol.* **75**, 693–699.

Polson, A., and von Wechmar, M. B. (1980). *J. Gen. Virol.* **52**, 179–181.

Puchta, H., Ramm, K., and Sänger, H. L. (1988). *Nucleic Acids Res.* **16**, 4197–4216.

Raccah, B., Loebenstein, G., and Bar-Joseph, M. (1976). *Phytopathology* **66**, 1102–1104.

Randles, J. W. (1975). *Phytopathology* **65**, 163–167.

Riechmann, J. L., Laín, S., and García, J. A. (1990). *Virology* **177**, 710–716.

Rigden, J. E., and Rezaian, M. A. (1992). *Virology* **186**, 201–206.

Rizzo, T. M., and Palukaitis, P. (1990). *Mol. Gen. Genet.* **222**, 249–256.

Rochester, D. E., Kositratana, W., and Beachy, R. N. (1990). *Virology* **178**, 520–526.

Rochon, D. M., and Johnston, J. C. (1991). *Virology* **181**, 656–665.

Rochow, W. F., and Brakke, M. K. (1964). *Virology* **24**, 310–322.

Rochow, W. F., Blizzard, J. W., Muller, I., and Waterworth, H. E. (1976). *Phytopathology* **66**, 534–536.

Romaine, C. P., and Horst, R. K. (1975). *Virology* **64**, 86–95.

Rosner, A., and Bar-Joseph, M. (1984). *Virology* **139**, 189–193.

Sasaki, M., and Shikata, E. (1977). *Proc. Jpn. Acad., Ser. B* **53**, 109–112.

Schumacher, J., Meyer, N., Weidemann, H. L., and Riesner, D. (1986). *J. Phytopathol.* **115**, 332–343.

Semancik, J. S., and Weathers, L. G. (1972). *Virology* **47**, 456–466.

Shepherd, R. J., Bruening, E. G., and Wakeman, R. J. (1970). *Virology* **71**, 339–347.

Shikata, E., Sano, T., and Uyeda, I. (1984). *Proc. Jpn. Acad., Ser. B* 202–205.

Silber, G., and Burk, L. G. (1965). *Nature (London)* **206**, 740–741.

Singh, R. P. (1970). *Am. Potato J.* **47**, 225–227.

Singh, R. P., and Bagnall, R. H. (1968). *Phytopathology* **58**, 696–699.

Singh, R. P., and Boucher, A. (1987). *Phytopathology* **77**, 1588–1591.

Singh, R. P., and Boucher, A. (1991). *Plant Dis.* **75**, 184–187.

Singh, R. P., and Finnie, R. E. (1977). *Phytopathology* **67**, 283–286.

Stanley, J. (1983). *Nature (London)* **305**, 643–645.

Tabler, M., and Sänger, H. L. (1984). *EMBO J.* **3**, 3055–3062.

Taylor, C. E. (1972). *In* "Principles and Techniques in Plant Virology" (C. I. Kado and H. O. Agrawal, eds.), pp. 226–247. Van Nostrand-Reinhold, New York.

Teakle, D. S. (1972). *In* "Principles and Techniques in Plant Virology" (C. I. Kado and H. O. Agrawal, eds.), pp. 248–266. Van Nostrand-Reinhold, New York.

Thomas, W., and Mohamed, N. A. (1979). *Aust. Plant Pathol. Newsl.* **8**, 1–3.

Timoshenko, N. A., Vnuchkova, V. A., Vishnichenko, V. K., and Zavriev, S. K. (1989). *Sov. Agric. Sci.* **11**, 12–16.

Valverde, R. A., Dodds, J. A., and Heide, J. A. (1986). *Phytopathology* **76**, 459–465.

Van der Want, J. P. H., Boerjan, M. L., and Peters, D. (1975). *Neth. J. Plant Pathol.* **81**, 205–216.

Van Dorst, H. J. M., and Peters, D. (1974). *Neth. J. Plant Pathol.* **80**, 85–96.

Van Regenmortel, M. H. V. (1982). "Serology and Immunochemistry of Plant Viruses." Academic Press, New York.

Visvader, J. E., Forster, A. C., and Symons, R. H. (1985). *Nucleic Acids Res.* **13**, 5843–5856.

Vos, P., Jaegle, M., Wellink, J., Verver, J., Eggen, R., Van Kammen, A., and Goldbach, R. (1988). *Virology* **165**, 33–41.

Vunsh, R., Rosner, A., and Stein, A. (1990). *Ann. Appl. Biol.* **117**, 561–569.

Walter, B. (1981). *C. R. Seances Acad. Sci., Ser. 3* **292**, 537–542.

Ward, A., Etessami, P., and Stanley, J. (1988). *EMBO J.* **7**, 1583–1587.

Watson, M. A. (1972). *In* "Principles and Techniques in Plant Virology" (C. I. Kado and H. O. Agrawal, eds.), pp. 131–187. Van Nostrand-Reinhold, New York.

Weiland, J. J., and Dreher, T. W. (1989). *Nucleic Acids Res.* **17**, 4675–4687.

Winter, S., and Nienhaus, F. (1989). *Eur. J. For. Pathol.* **19**, 111–118.

Withers, L. A. (1991). *In* "Maintenance of Microorganisms and Cultured Cells" (B. E. Kirsop and A. Doyle, eds.), 2nd ed., pp. 243–267. Academic Press, San Diego, CA.

Worley, J. F., and Schneider, I. R. (1966). *Phytopathology* **56**, 1327.

Xiong, Z., and Lommel, S. A. (1991). *Virology* **182**, 388–392.

Young, M. J., Larkin, K. P. J., Waterhouse, P. M., and Gerlach, W. L. (1991). *Virology* **180**, 372–379.

Zaitlin, M., and Israel, H. W. (1975). *CMI/AAB Description Plant Viruses* No. 151.

Zaumeyer, W. J. (1962). *Phytopathology* **52**, 486.

Ziegler-Graff, V., Bouzouboa, S., Jupin, I., Guilley, H., Jonard, G., and Richards, K. (1988). *J. Gen. Virol.* **69**, 2347–2357.

CHAPTER 12

Characterization of Cultures Used for Biotechnology and Industry

Angela Belt

The previous 10 chapters have covered in some detail the current reliable methods to maintain cultures used by researchers in biotechnology and industry. The methods vary, depending on the inherent nature of the culture and the amount of long-term preservation research that has been conducted for each group of organisms or cultures. The overall intent or purpose, however, is the same for all cultures, i.e., their maintenance in stable form.

Depending on one's interest in the cultures being studied, "stable form" may have different meanings. Often the measure of the viability of a culture is considered a measure of the success of the preservation method. Although the importance of viability should not be minimized, in industry—where large investments of man-hours and resources have been made to discover, produce, or improve cultures that yield chemical commodity, antibiotic, enzyme, or other biomolecule—maintaining the viability alone may not be enough.

Cultures that survive the preservation method used may lose certain physiological and/or morphological characteristics. For example, under some lyophilization protocols, *Clostridium botulinum* Type E lose their ability to produce toxin (Shannon *et al.*, 1975), some *Salmonella* lose somatic antigens, and many antibiotic producers lose their ability to produce antibiotic (Dietz, 1975), even though they maintain viability. For a company involved in the production of pharmaceuticals or diagnostic products, such losses can be extremely serious. Genetic engineering firms may have concern over preservation methods that result in an increase in the population of mutants in a given culture (Starr *et al.,* 1981; Ashwood-Smith and Grant, 1976; Ashwood-Smith, 1965), while viability remains, or in loss of plasmid stability in an otherwise viable host.

Given these concerns, choosing which maintenance or preservation method to use should be evaluated in terms of maintaining the characteristics of greatest interest or value to the researcher. Therefore, one should (1) determine the most important or desirable characteristics of a given culture, (2) develop a quantitative method or assay to evaluate these characteristics, and (3) establish a characterization baseline from which to evaluate the resulting effects of preserving the culture.

Desirable Characteristics of Cultures

It may be presumptious to suggest that some characteristics of a bacterium, for example, are of more value or worth than others. In nature, the sum of all of an organism's genetic makeup, expressed and nonexpressed, contributes to its successful existence and reproduction. And in an ideal research situation, one would have unlimited resources with which to explore cultures in every minute detail, seeking to understand and preserve every identifiable trait, no matter how obsure or seemingly minor. However, economic necessity usually requires that priorities be established and tasks be completed. The objective of this chapter, therefore, is to offer suggestions as to which characteristics may need to be evaluated, and how that can be done. How much weight these suggestions are given, and whether and how they are incorporated into the reader's laboratory, will depend on the objectives and work situation of each researcher.

The first step to evaluating the importance of the characteristics of a culture is to ask why the culture is of importance and why it needs to be maintained. If one is studying the taxonomy of plant pathogenic fungi, for instance, it is crucial to retain the morphological characteristics of the original isolate in as close to a "wild-type" state as possible, and to minimize strain deterioration that might result from repeated subculturing. On the other hand, if one has invested time developing mutants with higher yields of a given metabolite, then preventing reversion to an original, less productive culture is paramount. If one has cloned *Escherichia coli,* containing plasmids that produce IL-2, then the yield of IL-2 is obviously of greater importance than any characteristics the culture may share with nonrecombinant *E. coli.* Tissue culture cells used as quality controls for *Mycoplasma* testing of production lines must themselves be guaranteed free of *Mycoplasma.* And recombinant host strains chosen for certain genetic markers, e.g., λ^-, amber suppresser, or proline requirement, must be evaluated periodically in terms of those markers. Bearing these considerations in mind, Table 1 provides a list of suggested characteristics to evaluate. It is important to evaluate the culture before preserving in order to establish a baseline level of expression, as well as after preserving and on a routine basis.

Quantification

Although many of the traits listed in Table 1 are qualitative, e.g., cell appearance and pigmentation, the stability of cultures should be evaluated quantitatively whenever possible. This is especially important in the long term because staffing may change, and one person's interpretation of a "stable-looking" culture may differ from another's. In addition, acceptance levels must be established as a basis on which to reject or accept a culture's condition. Some methods and standards can be based on routine laboratory methods and are described below.

Culture Viability

Standard dilution plate counts can be used wherever applicable. A baseline count of colony-forming units/milliliter (cfu/ml) can be compared to cfu/ml when a culture is revived after preservation. Generally one seeks a method that results in a viability loss no greater than one log. Most cell death occurs during the initial steps of preservation (except for continuous subculture). Thus, plating out the culture within a week of processing can give a good idication of the success rate. Long-term losses can occur, however, perhaps because of equipment influence (e.g., breakdown of storage units), human influence (e.g., moving cultures from one freezer to another, with damaging temperature changes in the process), or through inherent sensitivity of the culture. It is advisable, therefore, to quantify viability each time a culture is subcultured or on a regular schedule. Cultures that cannot be dilution plated can be measured as hyphal extension rate for filamentous strains, microscopic cell counts, or other growth rates.

Culture Purity

Purity can be checked coincidentally with viability determinations. Generally, the standard for rejection of lots or preserved batches is anything less than 100% pure.

Product Formation

The method used to quantify the formation of a product will vary, depending on the metabolite of interest. If the metabolite is of particular

TABLE 1
Culture Characteristics to Evaluate

Characteristic	Example	Applications
Viability	Colony counting on plates if possible; otherwise, determine cell counts microscopically; for filamentous cultures, growth radius or hyphal extension over time	All cultures
Purity	Examine for contaminants through culturing and microscopy; PFLA/PCR, DNA hybridization, etc.	All cultures
Morphology	Culture appearance, e.g., color, odor, exudates, diffusible pigments, reverse, type and quantity of spores, occurrence of sectors	Fungi, algae, bacteria, actinomycetes, as applicable
	Cell appearance, fine structure, aberrant forms	Tissue culture cell
Tissue culture tests	Karyology, tumorigenicity, isoenzymology, species verification, antibody production	Cell cultures
Product formation	When applicable, measure product formation under defined conditions and quantify periodically (this may be expediated by development of standard methods performed by resident assay group, etc.)	All cultures
Recombinant traits	Host markers (e.g., thi⁻, F⁻, Supp)	Bacterial hosts
	Antibiotic resistance, plasmid size, restriction sites, sequence, location on chromosome (if inserted in eukaryotic host)	Plasmid-containing cultures
	Plaque formation efficiency, host specificity, plaque morphology	Bacteriophage

importance to the value of the culture, its production should be measured both prior and subsequent to preservation, under optimum conditions.

Recombinant Traits

PLASMIDS

There is an abundance of traits specific to recombinant cultures. The most obvious for recombinant bacteria is the presence or absence of plasmids. The most assured way of determining whether the plasmid is stable is sequencing, although this may be generally impractical on a routine basis. Plasmid isolation, restriction digests, and sizing can be performed quickly and routinely, using a sample of purified plasmid as the control. (The pure DNA can be maintained as a reference sample for long periods of time without breakdown through freezing at $-80°C$.)

In addition, although the plasmid construct may remain stable, it is important to know whether there is a demonstrable change in the percentage of the culture population maintaining plasmids. A simple and efficient method of determining this is the "patch test" (A. Belt, K. Bauer, J. Neway, unpublished). Cultures are plated onto agar without antibiotics, diluted such that a countable range is obtained. Next, individual colonies are picked with a sterile stick from the agar plate; the stick is touched first to another agar plate without antibiotic and then to one with antibiotic. After the colonies are "patched" and incubated, the percentage of colonies retaining plasmids can be determined by comparing the same colonies on each plate. Assuming a binomial distribution, the percentage of plasmid-bearing colonies can be determined as follows:

$$p^N = 1 - c,$$

where p is the fraction of plasmid-bearing colonies, N is the number of patches made, and c is the confidence limit (Caulcott et al., 1987). Within the confidence limits of the test (Figure 1), specifications for the acceptance or rejection of culture lots can be established. The procedure may be used to test a combination of traits as well—e.g., antibiotic resistance combined with carbohydrate fermentation on MacConkey plates.

Plates in eukaryote hosts, as above, can be checked for antibiotic resistance and product fromation. In addition to examining for the presence of plasmids, it is recommended that cultures be periodically examined to verify that any plasmids integrated into the host chromosome are intact and in the correct location. Gene replacements have been shown to lead to gross rearrangements of one or more chromosomes, undetected except by chromosome separation gels and Southern blots of these gels and restriction gels (MacKay, 1988). In addition, spontaneous chromosome rearrangement

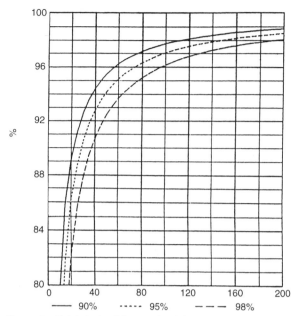

Figure 1. Patch test: minimum plasmid retention vs. number of patches.

in parent strains, which had not been subject to gene replacement efforts, were detected, emphasizing the need to characterize routinely not only the recombinant strain, but the parent or host as well (Caulcott, 1987).

BACTERIOPHAGE

Relatively little work has been done in the area of long-term preservation of bacteriophage. Workers often store the material over chloroform or at 5°C (Sambrook *et al.,* 1989), with subsequent loss of titer yearly. Freezing at ultralow temperatures has shown promising results for preservation of some bacteriophage (Clark *et al.,* 1962; Clark and Klein, 1966; Meyle and Kempf, 1964) as well as for lamdba libraries (Nierman *et al.,* 1987). In any event, in order to evaluate the efficacy of the preservation or maintenance method used, titers should be obtained before and after handling and other pertinent characteristics should also be examined. For example, plaque morphology, host specificity, and antiserum reaction rates may merit examination.

RECOMBINANT HOSTS

As for all other cultures, bacterial hosts should be thoroughly evaluated for their important characteristics, e.g., genetic markers, biochemical profiles, and serotyping. This is particularly important because many labs use a common source or "master stock" of hosts for numerous transformation experiments, which may eventually lead to construct–host combinations used in manufacturing or production lines. Cultures that are used to prepare competent cells may produce erratic competence and lower transformation efficiencies, depending on how the cells are stored (Hanahan *et al.*, 1991). Periodic evaluation of these characteristics would be well warranted.

Summary

In summary, the goal of maintaining cultures is to ensure stability of characteristics. To do so, one must establish a baseline or starting point to which further evaluations (after preservation methods are used) are compared. The most complete approach to this task would be to evaluate all measurable or observable characteristics. Evaluating stability of the cultures based on those features most important or relevant to the work being conducted is a minimum requirement.

References

Ashwood-Smith, M. J. (1965). Genetic stability of bacteria to freezing and thawing. *Cryobiology* **2**, 39–43.

Ashwood-Smith, M. J., and Grant, E. (1976). Mutation induction in bacteria by freeze-drying. *Cryobiology* **13**, 206–213.

Caulcott, C. A. (1987). Investigation of the effect of growth environment on the stability of low-copy-number plasmids in *Escherichia coli. J. Gen. Microbiol.* **133**, 1881–1889.

Clark, W. A., and Klein, A. (1966). The stability of bacteriophages in long term storage at liquid nitrogen temperatures. *Cryobiology* **3**, 68–75.

Clark, W. A., Horneland, W., and Klein, A. G. (1962). Attempts to freeze some bacteriophages at ultralow temperatures. *Appl. Microbiol.* **10**, 463–465.

Deitz, A. (1975). Nitrogen preservation of stock cultures of unicellular and filamentous microorganisms. Round Table Conference on the Cryogenic Preservation of Cell Cultures (A. P. Rinefret and B. LaSalle, eds.). National Academy of Science, Washington, D.C.

Hanahan, D., Jessee, J., and Bloom, F. R. (1991). Plasmid transformation of *Escherichia coli* and other bacteria. *In* "Methods in Enzymology," Vol. 204, pp. 63–113. Academic Press, New York.

MacKay, V., (1988). Genetic variation during development of yeast strains. USFCC Symposium: The Role of Culture Collections in the Study and Preservation of Biological Diversity. Davis, CA, August 17, 1988.

Meyle, J. S., and Kempf, J. E. (1964). Preservation of T_2 bacteriophage with liquid nitrogen. *Appl. Microbiol.* **12**, 400–402.

Nierman, W. C., Trypus, C., and Deaven, L. L. (1987). Preservation and stability of bacteriophage lambda libraries by freezing in liquid nitrogen. *Biofeedback* **5**(8), 724–727.

Sambrook, J., Fritsch, E. F., and Maniatis, T. (1989). "Molecular Cloning: A Laboratory Manual," 2nd Ed. Cold Spring Harbor Laboratory Press, Cold Spring Harbor, New York.

Shannon, J. E., Gherna, R. L., and Jong, S. C. (1975). The role of liquid nitrogen refrigeration at the American Type Culture Collection. Round Table Conference on the Cryogenic Preservation of Cell Cultures (A. P. Rinefret and B. LaSalle, eds.). National Academy of Science, Washington, D.C.

Starr, M. P., Stolp, H., Truper, H. G., Balows, A., and Schlegel, H. G., eds. (1981). "The Prokaryotes: A Handbook on Habitats, Isolation, and Identification of Bacteria." Springer-Verlag, Berlin and New York.

Index

Actinomycetes
  actinophage, 90
  characterization of cultures, 87–89
  classification, 86
  continuous subculture, 91
  cryopreservation
    straw, 94
    vials, 94–95
  culture conditions, 89–91
  diversity, 86
  dried blood preparations, 92–93
  industrial importance, 85–87
  lyophilization, 93–94
  preservation methods, 21, 89, 95
    batch testing, 96–97
    documentation, 97
    safety, 97
    security, 97
    selection, 96
  soil preparations, 92
  spore formation, 21
  viability and stability, evaluation,
    95–96
Agar
  cryoprotection, 118
  discovery as culture medium, 16–17
  industrial applications, 35–36

Algae
  cryopreservation, 37–38
    cold hardening, 40
    cooling rate, 40–41
    cryprotectants, 38
    culture age, 40
  diversity, 30
  economic importance, 31, 33–36
  international depositories, 45–60
  lineages, 30–31
  lyophilization, 41–42
  medical importance, 36–37
  prokaryotic forms, 29–30
  serial transfer
    culture facilities, 44
    culture media, 43–44
    history, 42–43
  taxonomy, 29–30, 32–33
Amino acid, microbial synthesis, 20
Animal cell culture, *see* Tissue culture
Antibiotics, history of production,
    19–20, 86
Aphid, plant virus vector, 234

Bacteria
  characterization methods, 69–71